Neural Networks and Micromechanics

Ernst Kussul • Tatiana Baidyk •
Donald C. Wunsch

Neural Networks and Micromechanics

Prof. Ernst Kussul
Center of Applied Research and Technological Development
Autonomous National University of Mexico (UNAM)
Mexico
ekussul@servidor.unam.mx

Prof. Tatiana Baidyk
Center of Applied Research and Technological Development
Autonomous National University of Mexico (UNAM)
Mexico
t.baydyk@ccadet.unam.mx

Prof. Donald C. Wunsch II
Dept. of Electrical and Computer Engineering
Missouri University of Science and Technology
Rolla, USA
dwunsch@mst.edu

ACM Computing Classification (1998): I.2, I.4, I.5, C.3

ISBN: 978-3-642-42611-7 ISBN: 978-3-642-02535-8 (eBook)
DOI 10.1007/978-3-642-02535-8
Springer Heidelberg Dordrecht London New York

Softcover re-print of the Hardcover 1st edition 2010

Cover design: KuenkelLopka, Heidelberg, Germany

Printed on acid-free paper

Springer is part of Springer Science+Business Media (www.springer.com)

Preface

Micromechanical manufacturing based on microequipment creates new possibilities in goods production. If microequipment sizes are comparable to the sizes of the microdevices to be produced, it is possible to decrease the cost of production drastically. The main components of the production cost - material, energy, space consumption, equipment, and maintenance - decrease with the scaling down of equipment sizes. To obtain really inexpensive production, labor costs must be reduced to almost zero. For this purpose, fully automated microfactories will be developed.

To create fully automated microfactories, we propose using artificial neural networks having different structures. The simplest perceptron-like neural network can be used at the lowest levels of microfactory control systems. Adaptive Critic Design, based on neural network models of the microfactory objects, can be used for manufacturing process optimization, while associative-projective neural networks and networks like ART could be used for the highest levels of control systems.

We have examined the performance of different neural networks in traditional image recognition tasks and in problems that appear in micromechanical manufacturing. We and our colleagues also have developed an approach to microequipment creation in the form of sequential generations. Each subsequent generation must be of a smaller size than the previous ones and must be made by previous generations. Prototypes of first-generation microequipment have been developed and assessed.

Interaction between neural networks and micromechanics does not have only one direction – while neural networks are helpful in micromechanics, micromechanics also can help to find new applications for neural networks. Currently, it is difficult to examine the effectiveness of neural networks in mechanical industry automation because each experiment in a mechanical factory is very expensive. Micromechanical factories will help us to examine different neural networks, compare them in mechanical production tasks, and recommend their use in conventional mechanics.

The results given in this book permit us to estimate optimistically the perspectives of neural network applications in micromechanics.

We have been working on neural networks, image recognition, and micromechanics for many years. Support from the National Academy of Sciences of the Ukraine, from the European Union (projects INTAS), from CONACYT and DGAPA (UNAM) in Mexico, from the US National Science Foundation the Mary K. Finley Endowment, the Missouri S&T Intelligent Systems Center, and from the company WACOM in Japan is gratefully acknowledged.

We want to thank our collaborators from the Ukraine and from Mexico for helping us bring this interesting project to fruition.

Finally, Ernst Kussul would like to thank his family; Tatiana Baidyk would like to thank her son Oleksandr Makeyev, her mother Mariya Baidyk and her sister Olga Malinina; and Don Wunsch would like to thank Hong and Donnie. Without their encouragement, understanding, and patience, this book would not exist.

E. Kussul, T. Baidyk, D. C. Wunsch

Contents

Chapter 1
Introduction

The title of the book, "Neural Networks and Micromechanics," seems artificial. However, the scientific and technological developments in recent decades demonstrate a very close connection between the two different areas of neural networks and micromechanics. The purpose of this book is to demonstrate this connection.

Some artificial intelligence (AI) methods, including neural networks, could be used to improve automation system performance in manufacturing processes. However, the implementation of these AI methods within industry is rather slow because of the high cost of conducting experiments using conventional manufacturing and AI systems. To lower the cost, we have developed special micromechanical equipment that is similar to conventional mechanical equipment but of much smaller size and therefore of lower cost. This equipment could be used to evaluate different AI methods in an easy and inexpensive way. The proved methods could be transferred to industry through appropriate scaling. In this book, we describe the prototypes of low cost microequipment for manufacturing processes and the implementation of some AI methods to increase precision, such as computer vision systems based on neural networks for microdevice assembly and genetic algorithms for microequipment characterization and the increase of microequipment precision.

The development of AI technologies opens an opportunity to use them not only for conventional applications (expert systems, intelligent data bases [1], technical diagnostics [2, 3] etc.), but also for total automation of mechanical manufacturing. Such AI methods as adaptive critic design [4, 5], neural network-based computer vision systems [6–10], etc. could be used to solve automation problems. To examine this opportunity, it is necessary to create an experimental factory with fully-automated manufacturing processes. This is a very difficult and expensive task.

To make very small mechanical microequipment, a new technology was proposed [11–14]. This technology is based on micromachine tools and microassembly devices, which can be produced as sequential generations of microequipment. Each generation should include equipment (machine tools, manipulators, assembly devices, measuring instruments, etc.) sufficient for manufacturing identical but smaller equipment. Each subsequent equipment generation could be produced by

E. Kussul et al. *Neural Networks and Micromechanics*,
DOI 10.1007/978-3-642-02535-8_1, © Springer-Verlag Berlin Heidelberg 2010

the preceding one. The size of each subsequent generation's equipment is smaller than the overall size of the preceding generation.

The first-generation microequipment can be produced by conventional large-scale equipment. Using microequipment of this first generation, a second micro-equipment generation having smaller overall sizes can be produced.

We call this approach to mechanical microdevice manufacturing Microequipment Technology (MET) [15].

The proposed MET technology has many advantages:

(1) The equipment miniaturization leads to decreasing the occupied space as well as energy consumption, and, therefore, the cost of the products.
(2) The labor costs are bound to decrease due to the reduction of maintenance costs and a higher level of automation expected in MET.
(3) Miniaturization of equipment by MET results in a decrease of its cost. This is a consequence of the fact that microequipment itself becomes the object of MET. The realization of universal microequipment that is capable of extended reproduction of itself will allow the manufacture of low-cost microequipment in a few reproductive acts because of the lower consumption of' materials, energy, labor, and space in MET. Thus, the miniaturization of equipment opens the way to a drastic decrease in the unit cost of individual processing.

At a lower unit cost of individual micromachining, the most natural way to achieve high throughput is to parallelize the processes of individual machining by concurrent use of a great quantity of the same kind of microequipment. Exploitation of that great number of microsized machine tools is only feasible with their automatic operation and a highly automated control of the microfactory as a whole. We expect that many useful and proved concepts, ideas, and techniques of automation can be borrowed from mechanical engineering. They vary from the principles of factory automation (FMS and CAM) to the ideas of unified containers, clamping devices, and techniques of numerical control. However, the automation of micromanufacturing has peculiarities that will require the special methods of artificial intelligence.

Let us consider a general hierarchical structure of the automatic control system for a micromechanical factory. The lowest (first) level of the system controls the micromechanical equipment (the micromachine tools and assembly manipulators) and provides the simplest microequipment diagnostics and the final measurement and testing of production. The second level of the control system controls the devices that transport workpieces, tools, parts, and all equipment items; coordinates the operation of the lowest level devices; and provides the intermediate quality inspection of production and the more advanced diagnostics of equipment condition. The third control level contains the system for the automatic choice of process modes and routes for machining parts. The top (fourth) level of the control system detects non-standard and alarm situations and makes decisions regarding these situations, including communication with the operator.

We proceed from the assumption that no more than one operator will manage the microfactory. This means that almost all the problems arising at any control level during the production process should be solved automatically and that the operator must solve only a few problems, those that are too complex or unusual to be solved automatically. Since any production process is affected by various disturbances, the control system should be an adaptive one. Moreover, it should be self-learning because it is impossible to foresee all kinds of disturbances in advance. AI that is able to construct the self-learning algorithms and to minimize the participation of the operator appears especially useful for this task. AI includes different methods for creating autonomous control systems. The neural classifiers will be particularly useful at the lowest level of the control system. They could be used to select treatment modes, check cutting tool conditions, control assembly processes, etc. They allow for more flexibility in the control system. The system will automatically compensate for small deviations of production conditions, such as a change in the cutting tool's shape or external environment parameters, variations in the structure of workpiece materials, etc. AI will permit the design of self-learning classifiers and should provide the opportunity to exclude the participation of a human operator at this level of control.

At the second control level, the AI system should detect all deviations from the normal production process and make decisions about how to modify the process to compensate for the deviation. The compensation should be made by tuning the parameters of the lower-level control systems. Examples of such deviations are deviations from the production schedule, failures in some devices, and off-standard production. At this level, the AI system should contain the structures in which the interrelations of production process constituents are represented. As in the previous case, it is desirable to have the algorithms working without the supervisor.

The third control level is connected basically with the change of nomenclature or volume of the production manufactured by the factory. It is convenient to develop such a system so that the set-up costs for a new production or the costs to change the production volume are minimal. The self-learning AI structures formed at the lowest level could provide the basis for such changes of set-up by selection of the process parameters, the choice of equipment configuration for machining and assembly, etc. At the third control level, the AI structures should detect the similarity of new products with the products that were manufactured in the past. On the basis of this similarity, the proposals about the manufacturing schedule, process modes, routing, etc. will be automatically formed and then checked by the usual computational methods of computer aided manufacturing (CAM). The results of the check, as well as the subsequent information about the efficiency of decisions made at this level, may be used for improving the AI system.

The most complicated AI structures should be applied at the top control level. This AI system level must have the ability to reveal the recent unusual features in the production process, to evaluate the possible influence of these new features on the production process, and to make decisions about changing the control system parameters at the various hierarchical levels or for calling for the operator's help. At this level, the control system should contain the intelligence knowledge base, which can be created using the results of the operation of the lower-level control systems and knowledge from experts. At the beginning, expert knowledge of macromechanics may be used.

At present, many methods of AI are successfully used in industry [16, 17], Some of these also could be used for micromechanics. Though the problems of fully-automated factory creation cannot be investigated experimentally in conventional industry because of the high cost of the experiments, here we propose to develop a low-cost micromechanical test bed to solve these problems.

The first prototype of the first generation was designed and manufactured at the International Research and Training Centre of Information Technologies, which is a part of V. M. Glushkov Cybernetics Center, Ukraine.

The second prototype of the first generation microequipment was designed and examined in CCADET, UNAM. The prototypes use adaptive algorithms of the lowest level.

At present, more sophisticated algorithms based on neural networks and genetic algorithms are being developed. Below, we describe our experiments in the area of the development and applications of such algorithms.

This book is intended as a professional reference and also as a textbook for graduate students in science, engineering, and micromechanics. We expect it to be particularly interesting to computer scientists and applied mathematicians applying it to neural networks, artificial intelligence, image recognition, and adaptive control, among many other fields.

References

1. Eberhart R., Overview of Computational Intelligence and Biomedical Engineering Applications, Proceedings of the 20th Annual International Conference of the IEEE Engineering in Medicine and Biology Society, 3, 1998, pp. 1125–1129.
2. Hui T., Brown D., Haynes B., Xinxian Wang, Embedded E-diagnostic for Distributed Industrial Machinery, IEEE International Symposium on Computational Intelligence for Measurement Systems and Applications, 2003, pp. 156–161.
3. Awadallah M., Morcos M., Application of AI Tools in Fault Diagnosis of Electrical Machines and Drives. An Overview, *IEEE Transactions on Energy Conversion*, 18, Issue 2, 2003, pp. 245–251.
4. Werbos P., Advanced Forecasting Methods for Global Crisis Warning and Models of Intelligence. In: General Systems Yearbook, 22, 1977, pp. 25–38.
5. Prokhorov D., Wunsch D., Adaptive Critic Designs, *IEEE Transactions on Neural Networks*, 8, No. 5, 1997, pp. 997–1007.
6. Bottou L., Cortes C., Denker J., Drucker H., Guyon L., Jackel L., LeCun J., Muller U., Sackinger E., Simard P., Vapnik V., Comparison of Classifier Methods: A Case Study in Handwritten Digit Recognition. In: Proceedings of 12th IAPR International Conference on Pattern Recognition, 2, 1994, pp. 77–82.
7. Fukushima, K., Neocognitron: A Hierarchical Neural Network Capable of Visual Pattern Recognition, *Neural Networks*, 1, 1988, pp. 119–130.
8. Roska T., Rodriguez-Vazquez A., Toward Visual Microprocessors, Proceedings of the IEEE 90, Issue 7, July 2002, pp. 1244–1257.

9. Baidyk T., Application of Flat Image Recognition Technique for Automation of Micro Device Production, Proceedings of the International Conference on Advanced Intelligent Mechatronics AIM 2001, Italy, 2001, pp. 488–494.
10. Baidyk T., Kussul E., Makeyev O., Caballero A., Ruiz L., Carrera G., Velasco G., Flat Image Recognition in the Process of Microdevice Assembly, *Pattern Recognition Letters*, 25/1, 2004, pp. 107–118.
11. Kussul E., Micromechanics and the Perspectives of Neurocomputing. In: Neuron-like Networks and Neurocomputers, Kiev, Ukraine, 1993, pp. 66–75 (in Russian).
12. Kussul E., Rachkovskij D., Baidyk T., Talayev S., Micromechanical Engineering: A Basis for the Low-cost Manufacturing of Mechanical Microdevices Using Microequipment, *Journal of Micromechanics and Microengineering*, 6, 1996, pp. 410–425.
13. Kussul E.M., Rachkovskij D.A., Kasatkin A.M., Kasatkina L.M., Baidyk T.N., Lukovich V.V., Olshanikov V.S., Talayev S.A., Neural Network Applications in Micro Mechanics, Neural Systems of Information Processing, Kiev, 1996, Vol. 1, pp. 80–88 (in Russian).
14. Kussul E.M., Micromechanics as a New Area of Neural Network Applications. Proceedings of the 5th European Congress on Intelligent Techniques and Soft Computing. Aachen, Germany, 1997, Vol.1 pp. 521–523.
15. Kussul, E., Baidyk, T., Ruiz-Huerta, L., Caballero, A., Velasco, G., Kasatkina, L.: Development of Micromachine Tool Prototypes for Microfactories. *Journal of Micromechanics and Microengineering*, 12, 2002, pp. 795–812.
16. Wenxin Liu., Venayagamoorthy G., Wunsch D., A Heuristic Dynamic Programming Based Power System Stabilizer for a Turbogenerator in a Single Machine Power System. 39th IAS Annual Meeting on Industry Applications, 1, 2003, pp. 270–276.
17. Croce, F., Delfino, B., et al.: Operations and Management of the Electric System for Industrial Plants: An Expert System Prototype for Load-Shedding Operator Assistance. *IEEE Transactions on Industry Applications* 37, Issue 3, 2001, pp. 701–708.

Chapter 2
Classical Neural Networks

During the last few decades, neural networks have moved from theory to offering solutions for industrial and commercial problems. Many people are interested in neural networks from many different perspectives. Engineers use them to build practical systems to solve industrial problems. For example, neural networks can be used for the control of industrial processes.

There are many publications that relate to the neural network theme. Every year, tens or even hundreds of international conferences, symposiums, congresses, and seminars take place in the world. As an introduction to this theme we can recommend the books of Robert Hecht-Nielsen [1], Teuvo Kohonen [2], and Philip Wasserman [3], and a more advanced book that is oriented on the applications of neural networks and is edited by A. Browne [4]. In this book it is assumed that the reader has some previous knowledge of neural networks and an understanding of their basic mechanisms. In this section we want to present a very short introduction to neural networks and to highlight the most important moments in neural network development.

2.1 Neural Network History

Attempts to model the human brain appeared with the creation of the first computer. Neural network paradigms were used for sensor processing, pattern recognition, data analysis, control, etc. We analyze, in short, different approaches for neural network development.

2.2 McCulloch and Pitts Neural Networks

The paper of McCulloch and Pitts [5] was the first attempt to understand the functions of the nervous system. For explanation, they used very simple types of neural networks, and they formulated the following five assumptions according to the neuron operation:

E. Kussul et al., *Neural Networks and Micromechanics*,
DOI 10.1007/978-3-642-02535-8_2, © Springer-Verlag Berlin Heidelberg 2010

1. The activity of the neuron is an "all-or-none" process.
2. A certain fixed number of synapses must be excited within the period of latent addition in order to excite a neuron at any time, and this number is independent of previous activity and position of the neuron.
3. The only significant delay within the nervous system is synaptic delay.
4. The activity of any inhibitory synapse absolutely prevents excitation of the neuron at that time.
5. The structure of the net does not change with time.

The McCulloch-Pitts neuron is a binary device (two stable states of a neuron). Each neuron has a fixed threshold, and every neuron has excitatory and inhibitory synapses, which are inputs of the neuron. But if the inhibitory synapse is active, the neuron cannot turn on. If no inhibitory synapses are active, the neuron adds its synaptic inputs. If the sum exceeds or equals the threshold, the neuron becomes active. So the McCulloch-Pitts neuron performs simple threshold logic.

The central result of their paper is that the network of the simplest McCulloch-Pitts neurons can realize any finite complex logical expression and compute any arithmetic or logical function. It was the first connectionist model.

2.3 Hebb Theory

Hebb tried to work out the general theory of behavior [6]. The problem of understanding behavior is the problem of understanding the total action of the nervous system, and vice versa. He attempted to bridge the gap between neurophysiology and psychology. Perception, learning in perception, and assembly formation were the main themes in his scientific investigations. Experiments had shown perceptual generalization. The repeated stimulation of specific receptors will lead to the formation of an "assembly" of association-area cells which can act briefly as a closed system. The synaptic connections between neurons become well-developed. Every assembly corresponds to any image or any concept. The idea that an image is presented by not just one neuron but by an assembly is fruitful. Any concept may have different meanings. Its content may vary depending on the context. Only the central core of the concept whose activity may dominate in the system as a whole can be almost unchangeable. The possible presentation of an image or concept with one neuron deprives this concept of its features and characteristics. The presentation with a neuron assembly makes possible a concept or image description with all features and characteristics. These features can be influenced by the context of the situation where the concept is used. For example, we create the model of the concept "building" (Fig. 2.1). We can observe the building from different positions. A perceived object (building) consists of a number of perceptual elements. We can see many windows or a door.

But from different positions there are walls and a roof of this building. In an assembly that is the model of the concept "building," a set of neurons corresponds

Fig. 2.1 Model of the concept "building"

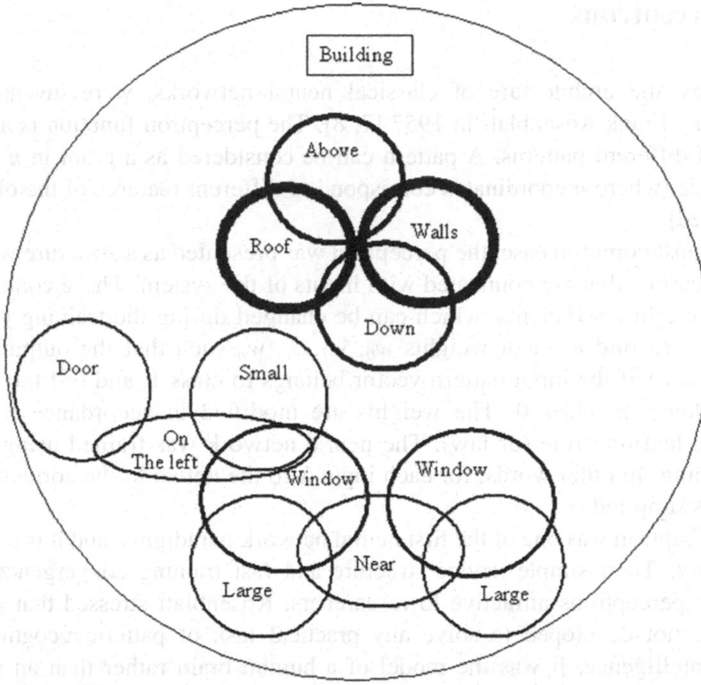

Fig. 2.2 Neural presentation of a model "building"

to the walls, other neurons correspond to windows, and others correspond to the white color of the walls, and so on. The more frequently perceived features of the building form the core of the assembly, and rare features create a fringe of the assembly (Fig. 2.2).

Due to the fringe of the assembly, different concepts may have a large number of associations with other concepts. "Fringe" systems were introduced by Hebb to explain how associations are provided. Different circumstances lead to varying fringe activity. If it is day, the white color of the building will be observed, and in the model the neuron set that corresponds to color will be excited. "Core" is the most connected part of the assembly. In our example, the core will be neurons that correspond to walls and windows. The conceptual activity that can be aroused with limited stimulation must have its organized core, but it may also have a fringe content, or meaning, that varies with the circumstances of arousal.

An individual cell or neuron set may enter into more than one assembly at different times. The single assembly or small group of assemblies can be repeatedly aroused when some other activity intervenes. In vision, for example, the perception of vertical lines must occur thousands of times an hour; in conversation, the word "the" must be perceived and uttered with very high frequency; and so on.

2.4 Perceptrons

Perceptrons, the architecture of classical neural networks, were invented and proposed by Frank Rosenblatt in 1957 [7, 8]. The perceptron function is a classification of different patterns. A pattern can be considered as a point in n-dimensional space (where n coordinates correspond to different features of the object to be classified).

In the most common case, the perceptron was presented as a structure with one layer of neurons that are connected with inputs of the system. These connections have the weight coefficients, which can be changed during the training process. The goal is to find a set of weights $w_0, w_1, \ldots, w_n$ such that the output of the perceptron is 1 if the input pattern vector belongs to class 1, and 0 if the pattern vector belongs to class 0. The weights are modified in accordance with the perceptron learning rule (or law). The neural network was trained using supervised training. In other words, for each input X to the network, the correct output Y also was supplied.

The perceptron was one of the first neural network paradigms, and it is still used occasionally. Their simple device structure and fast training convergence made Rosenblatt perceptrons attractive to researchers. Rosenblatt stressed that perceptrons were not developed to solve any practical task of pattern recognition or artificial intelligence. It was the model of a human brain rather than an applied technical device. However, it was clear that perceptrons could be used in practical applications, too.

Often, the Rosenblatt perceptron is considered a one-layer perceptron [9, 10]. Three-layered Rosenblatt perceptrons usually are mentioned in an historical context [11], though Rosenblatt investigated mainly three-layered perceptrons. It is interesting to build new classifiers on the base of the three-layered Rosenblatt perceptron and examine whether they can compete with the modern neural classifiers.

Analyzing the principal deficiencies of perceptrons, Rosenblatt mentioned the following [8]:

1. An excessively large system may be required.
2. The learning time may be excessive.
3. The system may be excessively dependent on external evaluation during learning.

4. The generalization ability is insufficient.
5. The ability to separate essential parts in a complex sensory field (analytic ability) is insufficient.

These points should be revised in the context of modern computer capabilities. Currently, computers cannot implement neural networks comparable with the human brain, which contains many billions of neurons, but it is possible to simulate the neuron structures containing up to, and in some cases larger than, a million neurons. In this case, it is interesting to know how the number of associative neurons influences Rosenblatt perceptron performance.

We studied and described several modifications of Rosenblatt perceptrons and experiments with them (O. Makeyev participated in this investigation) [12–19]. These experiments show that it is possible to overcome the above-mentioned problems using modern hardware. In the experiments, the number of associative neurons was changed from 1,000 to 512,000. The proposed perceptrons were tested on a benchmark MNIST data set for handwritten digits recognition [20, 21]. The performance of the modified Rosenblatt perceptron, having 512,000 neurons, is 99.2% on this database. As computer technology improves, larger capacity recognizers become feasible and higher recognition rates become possible. There are data about different classifiers' performances on this database. The best classifier on this database shows 99.3% [21].

Bernard Widrow was working along similar lines using systems known as Adalines (ADAptive LINear Element - a single processing unit with threshold non-linearity) [22, 23]. Widrow, along with his graduate student M. Hoff, proposed the Widrow/Hoff learning law or delta rule for neural network training. In Widrow learning, the goal is to find the best possible weight vector (for a very simple type of processing element) in terms of a least mean squared error performance function criterion. This learning rule is one of the most powerful and guarantees finding this optimum weight vector from any starting point.

But in 1969, Marvin Minsky and Seymour Papert attacked neural network research. They used predicates to describe the perceptron work [24]. In particular, the following points are critical remarks concerning perceptron functioning:

1. The idea of thinking about classes of geometrical objects as classes of n-dimensional vectors $(a_1, \ldots, a_n)$ loses the geometric individuality of the patterns and leads only to a theory that can do little more than *count* the number of predicates.
2. Little attention has been paid to the size, or more precisely, the information content, of the parameters $(a_1, \ldots, a_n)$. Some examples exist where the ratio of the largest to the smallest of the coefficients is meaninglessly large. In some cases, the information capacity needed to store $a_1, \ldots, a_n$ is even greater than that needed to store the whole class of figures defined by the pattern.
3. Closely related to the previous point is the problem of time of convergence in a learning process.

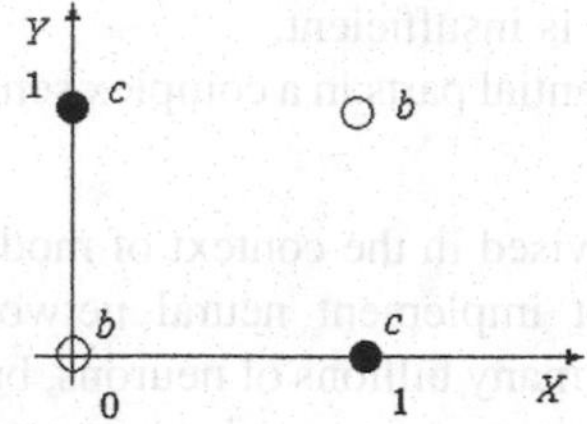

Fig. 2.3 Example of EXCLUSIVE OR (XOR) classification problem

Minsky and Seymour pointed out that a single-layer network can only classify data sets that are linearly separable and hence cannot solve problems such as the EXCLUSIVE OR (XOR) (Fig. 2.3). The input vectors (0, 0) and (1, 1) belong to class *b*, while the input vectors (0, 1) and (1, 0) belong to class *c*. It is clear that there is no linear decision boundary that can classify all four points correctly. This problem can be generalized to *n* dimensions, when it is known as the *n*-bit parity problem.

This was the sunset of neural network investigation. Artificial intelligence methods were developed and took the place of neural network investigation. From 1969 until 1982, neural network investigations had to go underground in the US, but in the Soviet Union, Europe, and Japan these investigations continued [25–30]. For example, in 1972–1975, the first autonomous transport robot was created in the USSR [26, 27]. The robot demonstrated obstacle avoidance and purposive movement in a natural environment. TAIR was a three-wheel power barrow equipped with a system of sensors as rangefinder and tactile sensors. It was controlled by a hardware-implemented neural network (the network nodes were realized as special transistor electronic circuits; connections between nodes were realized as resistors). TAIR is presented in Fig. 2.4.

While in motion, the robot was supposed to avoid obstacles such as people, trees, park benches, and so on. Coordinates of a point in the environment gave the target of the robot's motion. It was evident from the behavior of TAIR that, in principle, it is possible to create an entirely autonomous robot operated by a hardware-implemented neural network. At the same time, it showed the overall complexity of organization of the robot's interaction with the natural environment, as well as the necessity of using trainable neural networks.

2.5 Neural Networks of the 1980s

In the early 1980s, a new wave of interest arose due to the publication of John Hopfield [31], a researcher in the field of biophysics. He described the analogy between Hebb's neural network model and the certain class of physical systems. His efforts allowed hundreds of highly qualified scientists and engineers to join in

Fig. 2.4 The robot TAIR was controlled by a hardware-implemented neural network

the neural network investigation. At this time, the DARPA (Defense Advanced Research Projects Agency) project was initiated.

Around 1986, the new term "neurocomputer" appeared. Many international conferences on neural networks, neurocomputing, and neurocomputers took place all over the world. Hundreds of firms dedicated to neural network technology development and production were established. For example, the neurocomputer Mark III was built at TRW, Inc. during 1984–1985, followed by Mark IV [1]. In 1988, the firm HNC (Hecht-Nielson Corporation) produced the neurocomputer "ANZA plus," which can work together with PC 386, Sun. In the same year, the neurocomputer Delta II was produced by the firm SAIC.

In the department of network system of information processing, at the Institute of Cybernetics, Kiev, Ukraine, the first neurocomputer "NIC" was created in 1988–1989 [32, 33] under the direction of Ernst Kussul. This neurocomputer is presented in Fig. 2.5. It was built on a domestic element base and was a personal computer add-on. Kussul put forward and analyzed a new neural network paradigm, which enabled the creation of neuron-like structures. These structures are known as associative-projective neuron-like networks [34–36].

After that, in 1991–1992, the Ukrainian-Japanese team created a new neurocomputer that used a more advanced element base. It was named "B-512," and it is presented in Fig. 2.6. Kussul and his collaborators and disciples Tatiana Baidyk, Dmitrij Rachkovskij, Mikhail Kussul, and Sergei Artykutsa participated in the neurocomputer development together with the Japanese investigators from "WACOM," Sadao Yamomoto, Masao Kumagishi, and Yuji Katsurahira.

Fig. 2.5 First neurocomputer "NIC" developed at the Institute of Cybernetics, Kiev, Ukraine

Fig. 2.6 The neurocomputer "B-512" was created in 1992

The latest neurocomputer version was developed and tested on image recognition tasks. For example, the task of handwritten words recognition was resolved on this neurocomputer [37].

2.6 Modern Applications of Neural Network Paradigms

There are different approaches to neural network presentation, and different paradigms of neural networks have been developed. Among them, the most popular have been the Hopfield neural network [38, 39], adaptive resonance theory developed by S. Grossberg and G. Carpenter [40, 41], Kohonen neural networks [42], Fukushima cognitron and neocognitron [43], backpropagation [44–46], and adaptive critic design [47, 48].

2.6.1 Hopfield Neural Networks

Hopfield described his neural networks in 1982 [38]. The structure of this neural network has neural processing elements. The output of every processing element is the input of other neural processing elements. The transfer function of every processing element is:

$$X_i^{t+1} = \begin{cases} 1, & if \quad \sum_{j=1}^{n} w_{ij} X_j^t > T_i, \\ X_j^t, & if \quad \sum_{j=1}^{n} w_{ij} X_j^t = T_i, \\ -1, & if \quad \sum_{j=1}^{n} w_{ij} X_j^t < T_i, \end{cases} \tag{2.1}$$

for $i = 1, \ldots, n$,
where w_{ij} is the weight of the input with the restrictions $w_{ij} = w_{ji}$ and $w_{ii} = 0$, and T_i is the threshold. The behavior of the Hopfield network is organized in such a way as to minimize the energy function. No matter what its initial state, the Hopfield network always converges to a stable state in a finite number of processing element update steps.

There are many developments of the Hopfield network which are used in different applications. For example, the Hopfield network is a base of a hybrid Hopfield network-simulated annealing algorithm used for frequency assignment in satellite communications systems [49]. They use a fast digital Hopfield neural network, which manages the problem constraints, hybridized with simulated annealing, which improves the quality of the solutions obtained. Another example is a Hopfield neural network application for general predictive control [50]. In this case, the Hopfield neural network is used to solve quadratic optimizing problems. Existing predictive control methods are very complex and time consuming. With this proposition of a Hopfield neural network application and the development of a neural network chip, the method has a promising future in industry. The Hopfield neural network is used in information retrieval systems, too [51]. In recent years, with the rapid development of the Internet and easy access to a large amount of information on it, information retrieval has become more indispensable for picking out useful data from the massive resources. With the heuristic function of the Hopfield network, this model is used in query expansion, and therefore can solve the problems of information overload and word mismatch to some extent.

2.6.2 Adaptive Resonance Theory (ART)

Stephen Grossberg and Gail Carpenter [40, 41, 52] developed and introduced a variety of neural network paradigms. The most popular is adaptive resonance theory, which finds an application for different task solutions.

For example, driving safety is a very important consideration for the automotive industry and for consumers. The methods for improving driving safety can be roughly categorized into passive or active. Passive means (e.g., seatbelts, airbags, etc.), which have significantly reduced traffic fatalities, were originally introduced to diminish the degree of injury during an accident. Active means, on the other hand, are designed to prevent accidents in the first place. A driver assistance system is one kind of active system that is intended to alert a driver to the potential of a dangerous situation as soon as possible. Detecting critical changes in the driving environment is an important task in driver assistance systems. A computational model was developed for this purpose [53]. This model includes three components: sensory, perceptual, and conceptual analyzers. They use visual sensors (cameras and video camcoders). Each video sequence was down-sampled to a frame rate of 5 Hz to reduce the processing load on the computer. This frame rate is also fast enough for a driver to respond to any of the environmental changes. The size of each input image (320×240 pixels) was reduced to 160 x 120 pixels to reduce the processing time. A number of video sequences were collected and categorized into seven classes, referred to as the right-lane-change, left-lane-change, tunnel entry, tunnel exit, freeway entry, freeway exit, and overpass ahead. Each class was further divided into two groups, termed the "day" and "night" groups. A vision system for detecting critical changes in driving was developed. In this framework, both temporal and spatial information are extracted from input video sequences. The extracted information serves as input stimuli to a spatiotemporal attention neural network. The attention pattern associated with the focus, together with the location and direction of motion of the pattern, form what Chiung-Yao Fang et al. call a categorical feature. Thereafter, based on this feature, the class of the attention pattern and, in turn, the change in driving environment corresponding to the class is determined using a configurable adaptive resonance theory (CART) neural network. This is the work of the conceptual analyzer. Various changes in the driving environment, both in the daytime and at night, have been tested [53].

Adaptive resonance theory was applied to resolve large-scale traveling salesman problems (TSP) [54]. TSP is a very difficult optimization issue in the field of operations research. Using adaptive resonance theory and local optimization to divide and conquer large scale TSP has the advantages of increased scalability and parallelism.

An ART-2 network was used to develop a strategy for the navigation of mobile robots in uncertain indoor environments [55]. More exactly, a modified ART-2 network was put forward to identify the surrounding environment correctly for mobile robots. Path planning is one of the vital tasks in the navigation of autonomous mobile robots and may be divided into two categories: global path planning based on a priori complete information about the environment and local path planning based on sensor information in uncertain environments where the size, shape, and location of obstacles are unknown. Local path planning could be called reactive strategies. The neural networks and fuzzy controllers have proved to perform well in reactive navigation applications. Computer simulations were made for design strategies of environment classifiers based on a modified ART-2 network and fuzzy controller. One more example of an ART application is connected with a mobile vehicle [56]. In this case, an ART network is used for image processing. On the robot's path are several labels, the letters L, R, B, F, and S, which represent turning left or right, moving backward or forward, and stopping. ART is used to recognize them and give out the signal to control a mobile vehicle. Other experiments were conducted involving obstacle avoidance. Eight obstacle patterns were selected to train the ART network. The potential to add new categories is very important for this type of task.

In another study [57], the ART family of neural networks was used to develop a speaker recognition system. This system consists of two modules: a wavelet-based feature extractor and a neural-network-based classifier. Performance of the system has been evaluated using the gender recognition and speaker recognition problems. In the gender recognition problem, the highest accuracy was 90.33%; in the speaker recognition problem, ART-based classifiers have demonstrated recognition accuracy of 81.4%.

2.6.3 Self-Organizing Feature Map (SOFM) Neural Networks

Teuvo Kohonen published his first articles in the seventies [29]. He applied a specific type of neural network – Self-Organizing Feature Map (SOFM) designs with the following self-organization training principles. The principal idea is that a set of processing elements arrange their weights in such a way that they are distributed in space with a density approximately proportional to the probability density of the input vector.

This approach found a place in modern investigations and applications. For example, a Kohonen self-organizing map can be used for unsupervised segmentation of single-channel magnetic resonance (MR) images [58]. The introduction of advanced medical techniques, such as MRI, has dramatically improved the quality of the diagnosis and treatment of brain pathologies. The image information in such systems is complex and has a great number of dimensions. The availability of segmentation techniques is useful in assisting medical experts in the diagnosis and treatment of tumors. The segmentation of single-channel magnetic resonance images is a daunting task due to the relatively small amount of information available at each pixel site. This method has been validated on both simulated and real images of volunteers and brain tumor patients. This is the first step in developing a fully automatic segmentation method.

The other area where Kohonen maps found an application is for a distributed measurement system for water quality monitoring [59]. Water quality monitoring of rivers and seas represents an important task of life-quality assessment. This monitoring is characterized by multi-parameter measurement capabilities. The main parameters associated with water quality inspection can be classified in three categories: physical, chemical, and biological parameters. They use an advanced processing of data sensors based on auto-associative neural networks (Kohonen maps) in order to offer a global water quality representation for a large monitored area.

One more interesting application of Kohonen maps is feature selection for object extraction [60]. Selecting a set of features that are optimal for a given task is a problem that plays an important role in pattern recognition, image understanding, and machine learning. Li Pan et al. used Kohonen maps for continuous data discretization in texture-recognition tasks. As the test task, they used tree recognition from aerial images.

Also, Kohonen maps were used for color image compression [61]. With the development of multimedia technology and the Internet, image communication including transmission, display, and storage at high speed has become increasingly important. In this case, the hardware design for a neural-network-based color image compression was developed. Compression using neural networks is advantageous due to their features such as inherent parallelism, regular topology, and their relatively small number of well-defined arithmetic operations involved in their learning algorithms. So, VLSI implementation of Kohonen's map neural network is well suited to color image compression due to its topological clustering property. In this case, similar colors are clustered together and can be represented by one color. The time complexity of the proposed scheme is linear in the image size. With ASIC implementation the compression time is only a few milliseconds for images of sizes up to 512×512 pixels.

Extended Kohonen networks were used for the pose control of microrobots in a nanohandling station [62]. These are only a few examples among many others.

2.6.4 Cognitron and Neocognitron

Kunihico Fukushima developed and proposed the *cognitron* neural network [30], which is an example of a hierarchical network [43]. It was initially proposed as a neural network model of the visual system that has a hierarchical multilayered architecture similar to the classical hypothesis of Hubel and Wiesel [63, 64]. The model consists of S- and C-cells. S-cells resemble simple cells of the primary visual cortex, and C-cells resemble complex cells. The layers of S- and C-cells are arranged alternately in a hierarchical structure. S-cells feature extracting cells and have variable input connections, which can be modifiable during training. C-cells have fixed, non-modifiable input connections.

In recent years, these ideas have found new applications [65–67], and [68]. In Japan, at Toyota Central R&D Labs., Inc., the neocognitron is used for position detection and vehicle recognition [65]. This system is tolerant to deformations and shifts in the position of a vehicle. K. Fukushima [66] applies the neocognitron for handwritten digit recognition, but several new ideas have been introduced, such as the inhibitory surround in the connections from S-cells to C-cells. Fukushima also applies the neocognitron to the recognition of patterns that are partly occluded [68]. Other authors [67] have modified the neocognitron and used it for breast cancer detection. The so-called Shape Cognitron (S-Cognitron) is composed of two modules and was introduced to classify clustered micro calcifications, which generally present an early sign of breast cancer. The first module serves as a shape orientation layer and converts first-order shape orientations into numeric values. The second module is made up of a feature formation layer followed by a probabilistic neural-network-based classification layer. This system was tested on the 40-mammogram database provided by the Department of Radiology at the University of Hospital Nijmegen in the Netherlands and showed promising results.

2.6.5 Backpropagation

The backpropagation method was introduced by Paul Werbos in 1974 [46]. In 1985 –1986, D. Rumelhart and others worked out and applied this mechanism for neural network training [44, 45].

At present, the backpropagation method is used to resolve different practical tasks. For example, it is used for retrieval algorithms for geophysical parameters in the retrieval of atmospheric water vapor and cloud liquid water content over oceans from brightness temperatures measured by the multi-frequency scanning microwave radiometer launched onboard satellite [69]. These studies have demonstrated the great potential of neural networks in a large variety of remote sensing and meteorological applications. For this concrete task, the multilayer perceptron (MLP) with backpropagation training algorithm was used. MLP has the ability to detect multiple nonlinear correlations from the training database. MLP has advantages over statistical regression methods [70].

The backpropagation method is used very often in medicine, for example, for classifying balance disorders using simulated sensory deficiency [71]. This task is important for medical rehabilitation of patients with sensory deficiency. Another example is connected with the classification of neck movement patterns related to Whiplash-associated disorders (WADs) using a resilient backpropagation neural network [72]. WADs are a common diagnosis after neck trauma, typically caused by rear-end car accidents. Neck movement was used as input. Rotation angle and velocity were calculated. A principal component analysis was performed in order to reduce data and improve the backpropagation neural network performance. This method showed a predictivity of 0.89, which is a very promising result. The neck movement analysis combined with a neural network could build the basis of a decision support system for classifying suspected WADs.

Another interesting use is detecting sky in photographic images [73], which was done by collaborators of the Imaging Science Technology Laboratory, Kodak Research Laboratories, Eastman Kodak Company, USA. In their system, color classification is performed by a multilayer backpropagation neural network trained in a bootstrapping fashion to generate a belief map of sky color. Next, the region extraction algorithm automatically determines an appropriate threshold for the sky color belief map and extracts connected components. Finally, the sky signature validation algorithm determines the orientation of a candidate sky region using a physics-motivated model. With approximately half of the images containing blue sky regions, the detection rate is 96%. A feedforward neural network was structured with two hidden layers containing three and two neurons, and a single output neuron. The output of the network is a belief value between 0 and 1 for each pixel, 1 indicating a pixel highly likely to be blue sky.

In industry, the backpropagation neural network is used, for example, for fault diagnosis of circuit breakers [74]. The maintenance costs for aging circuit breakers (CBs) are significant. To reduce the cost, a condition-based maintenance is proposed. Many testing methods for characterizing the condition of a CB have been studied, such as contact travel time measurement and vibration analysis. In [74], wavelet packets are used to convert measured vibration data from healthy and defective CBs into wavelet features. Selected features highlighting the differences between healthy and faulty conditions are processed by a backpropagation neural network for classification. Testing has been done for three 66-kV CBs with simulated faults. Detection accuracy is shown to be far better than in other classical techniques such as the windowed Fourier transform, standalone artificial neural networks, or expert systems. The accuracy of detection for some defaults can be as high as 100%.

2.6.6 Adaptive Critic Design

One of the most interesting approaches to using adaptive networks to solve common problems in adaptive control and system identification is adaptive critic design [47, 48, 75]. In the broadest sense, this method is developed to handle large, noisy, nonlinear problems in real time. Many applications of neural networks to control were limited to the use of static or feedforward networks, adapted in relatively simple ways [75]. Many applications of that sort encountered problems such as limited application or slow adaptation.

Adaptive critic designs may be defined as designs that attempt to approximate dynamic programming in the general case. Dynamic programming is the method for finding an optimal strategy of action over time in a noisy, nonlinear environment [75]. The user supplies a utility function U and a stochastic model F of the environment to be controlled. Dynamic programming is used to solve for another function, J, which serves as a secondary or strategic utility function. The key theorem is that any strategy of action that maximizes J in the short term will also maximize the sum of U over all future times. Adaptive critic designs are defined more precisely as designs that include a critic network – a network whose output is an approximation to the J function, or to its derivatives, or to something very closely related to the two. There are different realizations of this approach; it is possible to say that an adaptive critic family of methods has developed.

For example, adaptive-critic-based neural networks have been used to design a controller for a benchmark problem in aircraft autolanding [76], to steer an agile missile [77], to build neurocontrollers for turbo generators in a multimachine power system [78], and to construct new learning methods (creative learning) for intelligent autonomous mobile robots [79].

We plan to use adaptive critic design in our future investigations of micromechanics and microfabrics control.

2.7 RTC, RSC, LIRA, and PCNC Neural Classifiers

We have worked out effective neural network classification systems. They have been developed since 1970 and used as control systems for mobile autonomous robots, in texture recognition tasks, in voice-based identity verification tasks, in handwriting and face recognition, and in the new micromechanics area. The most interesting neural classifiers are Random Threshold Classifier (RTC) [80, 81], Random Subspace Classifier (RSC) [12, 82], Neural Classifier LIRA (Limited Receptive Area) [12–19], and PCNC Neural Classifier [83, 84]. In this book, we describe all these models, obtain results, and summarize all of the advantages of our approach.

References

1. Hecht-Nielsen, R., Neurocomputing. Addison-Wesley, 1991, pp. 433.
2. Kohonen, T., Self-Organization and Associative Memory, Springer-Verlag, Berlin, 1984.
3. Wasserman, Philip D., Neural Computing: Theory and practice, Van Nostrand Reinhold, Inc., 1989.
4. Ed., Browne, A., Neural Network Analysis, Architectures and Applications. Institute of Physics Publishing, 1997, pp. 264.
5. McCulloch, W. S., and Pitts W. A., Logical Calculus of the Ideas Immanent in Nervous Activity. Bulletin of Math. Biophysics, 5, 1943, pp. 115–133.
6. Hebb, D. O., The Organization of Behavior. A Neuropsychological Theory. Wiley, New York, 1949, pp. 335.
7. Rosenblatt, F., "The perceptron: A Probabilistic Model for Information Storage and Organization in the Brain," Psychol. Rev., 65, pp. 386–408, 1958.

8. Rosenblatt, F., "Principles of Neurodynamics," Spartan Books, Washington, D.C., 1962, pp. 616.
9. Diggavi, S. N., Shynk, J. J., Engel, I., Bershad, N. J., "On the Convergence Behavior of Rosenblatt's Perceptron Learning Algorithm", *1992 Conference Record of the 26 Asilomar Conference on Signals, Systems and Computers*, Vol. 2, 1992, pp. 852–856.
10. Diggavi, S. N., Shynk, J. J., Bershad, N. J., Convergence Models for Rosenblatt's Perceptron Learning Algorithm, *IEEE Transactions on Signal Processing*, Vol. 43, Issue 7, July 1995, pp. 1696–1702.
11. Nagy, G., Neural Networks – Then and Now, *IEEE Transactions on Neural Networks*, Vol. 2, Issue 2, March 1991, pp. 316–318.
12. Kussul, E., Baidyk, T., Kasatkina, L., Lukovich, V., Rosenblatt Perceptrons for Handwritten Digit Recognition. Proceedings of International Joint Conference on Neural Networks "IJCNN'01", Washington, D.C., USA, July 15–19, 2001, pp. 1516–1521.
13. Kussul, E. M., Baidyk, T. N., Improved Method of Handwritten Digit Recognition Tested on MNIST Database, *15th International Conference on Vision Interface VI'2002*, Calgary, Canada, 2002, pp. 192–197.
14. Baidyk, T. N., Kussul, E. M., Application of Neural Classifier for Flat Image Recognition in the Process of Microdevice Assembly, International Joint Conference on Neural Networks, IEEE, USA, Hawaii, May 12–17, 2002 Vol. 1, pp.160-164.
15. Makeyev, O., Neural Interpolator for Image Recognition in the Process of Microdevice Assembly. In IEEE IJCNN'2003, Portland, Oregon, USA, July 20–24, 2003, Vol. 3, pp. 2222–2226.
16. Baidyk, T., Kussul, E., Makeyev, O., Image Recognition System for Microdevice Assembly, Twenty-First IASTED International Multi-Conference on APPLIED INFORMATICS AI2003. Innsbruck, Austria, pp. 243–248.
17. Baidyk, T., Kussul, E., Makeyev, O., Caballero, A., Ruiz, L., Carrera, G., Velasco, G., Flat Image Recognition in the Process of Microdevice Assembly. *Pattern Recognition Letters*, 2004, Vol. 25/1, pp. 107–118.
18. Toledo, G. K., Kussul, E., Baidyk, T., Neural Classifier LIRA for Recognition of Micro Work Pieces and their Positions in the Processes of Microassembly and Micromanufacturing. Proceedings of the 7 All-Ukrainian International Conference, 11–15 October 2004, Kiev, Ukraine, p. 17–20.
19. Kussul, E., Baidyk, T., Improved Method of Handwritten Digit Recognition Tested on MNIST Database. *Image and Vision Computing*, 2004, Vol. 22/12, pp. 971–981,
20. Bottou, L., Cortes, C., Denker, J.S., Drucker, H., Guyon, L., Jackel, L.D., LeCun, J., Muller, U.A., Sackinger, E., Simard, P., Vapnik, V., Comparison of Classifier Methods: A Case Study in Handwritten Digit Recognition, *Proceedings of 12th IAPR International Conference on Pattern Recognition*, 1994, Vol. 2, pp. 77–82.
21. LeCun, Y., Bottou, L., Bengio, Y., Haffner, P., Gradient-Based Learning Applied to Document Recognition, *Proceedings of the IEEE*, Vol. 86, No. 11, November 1998, pp. 2278–2344.
22. Widrow, B., Hoff, M., Adaptive Switching Circuits, 1960 IRE WESCON Convention Record, New York, 1960, 96–104.
23. Widrow, B., Generalization and Information Storage in Networks of ADALINE Neurons. In: Self-Organizing Systems, Ed. Yovitts G., Spartan Books, Washington DC, 1962.
24. Minsky, M., and Papert, S., Perceptrons. An Introduction to Computational Geometry. The MIT Press. 1969.
25. Amosov, N., Modeling of Thinking and the Mind. New York: Spartan Books, 1967, pp. 192.
26. Amosov, N., Kussul, E., Fomenko, V., Transport Robot with Network Control System. The 4th International Joint Conference on Artificial Intelligence. Tbilisi, Georgia, 1975, pp. 727–730.
27. Amosov, N., Kasatkin, A., Kasatkina, L., Active Semantic Networks in Robots with Independent Control. The 4th International Joint Conference on Artificial Intelligence. Tbilisi, Georgia, 1975, pp. 722–726.

28. Ivahnenko, A. G., Self-organizing Systems with Positive Feedback Loops. IEEE Transactions on Automatic Control, Vol. 8, Issue 3, July 1963, pp. 247–254.
29. Kohonen, T., Correlation Matrix Memories. IEEE Trans. Computers, C-21 (4), 1972, pp. 353–359.
30. Fukushima, K., Visual Feature Extraction by a Multilayered Network of Analog Threshold Elements. IEEE Trans. Systems. Sci.&Cyber., SSC-5 (4), 1969, pp. 322-333.
31. Hopfield, J., Neural Networks and Physical Systems with Emergent Collective Computational Abilities. Proc. Nat. Acad. Sci., USA, Vol. 79, 1982, pp. 2554–2558.
32. Kussul, E. M., Baidyk, T. N., Design of a Neural-Like Network Architecture for Recognition of Object Shapes in Images. Soviet Journal of Automation and Information Sciences (formerly Soviet Automatic Control), Vol. 23, No. 5, 1990, pp. 53–59.
33. Artykutsa, S. Y., Baidyk, T. N., Kussul, E. M., Rachkovskij, D. A., Textura Recognition with Neurocomputers. Preprint 91-8, Institute of Cibernetics of Academy of Science of Ukraine, 1991, pp. 20 (in Russian).
34. Kussul, E. M., Rachkovskij, D. A., Baidyk, T. N., Associative-Projective Neural Networks: Architecture, Implementation, Applications. Proc. of the Fourth Intern. Conf. "Neural Networks & their Applications", EC2 Publishing, Nimes, France, Nov. 4–8, 1991, pp. 463–476.
35. Kussul, E. M., Rachkovskij, D. A., Baidyk, T. N., On Image Texture Recognition by Associative-Projective Neurocomputer. Proc. of the ANNIE'91 conference. "Intelligent engineering systems through artificial neural networks", Ed. by C.H.Dagli, S. Kumara and Y.C. Shin, ASME Press, 1991, pp. 453–458.
36. Amosov, N., Baidyk, T., Goltsev, A., Kasatkin, A., Kasatkina, L., Kussul, E., Rachkovskij, D., Neurocomputers and Intelligent Robots. Ed. by N. Amosov, Kiev, Naukova Dumka, 1991, pp. 272 (in Russian).
37. Baidyk, T. N., and Kussul, E. M., Application of Genetic Algorithms to Optimization of Neuronet Recognition Devices, Cybernetics and System Analysis. Vol. 35, No. 5, 1999, pp. 700–707.
38. Hopfield, J., Neural Networks and Physical Systems with Emergent Collective Computational Abilities. Proc. Natl. Acad. Sci., 79, 1982, pp. 2554–2558.
39. Hopfield, J., Neurons with Grated Response Have Collective Computational Properties Like Those of Two-State Neurons. Proc. Natl. Acad. Sci., 81, 1984, pp. 3088–3092.
40. Grossberg, S., Studies of Mind and Brain: Neural Principles of Learning, Perception, Development, Cognition, and Motor Control. Reidel Press, Boston, 1982.
41. Carpenter, G., Grossberg, S., ART2: Self-organization of Stable Category Recognition Codes for Analog Input Patterns. Applied Optics, 26 (23), December 1987, pp. 4919–4930.
42. Kohonen, T., Self-organization and Associative Memory. Springer-Verlag, Berlin, 1984.
43. Fukushima, K., Neocognitron: A Hierarchical Neural Network Capable of Visual Pattern Recognition. Neural Networks, 1, 1988, pp. 119–130.
44. Rumelhart, D. E., Hinton, G. E., Williams, R., Learning Representations by Backpropagating Errors. Nature, 1986, 323, pp. 533–536.
45. Rumelhart, D., McClelland J. Parallel Distributed Processing: Explorations in the Microstructure of Cognition, I & II, MIT Press, Cambridge MA, 1986.
46. Werbos, P., Beyond regression: New Tools for Prediction and Analysis in the Behavioral Sciences. Doctoral Dissertation, Appl. Math., Harvard University, November 1974.
47. Werbos, P., Approximate Dynamic Programming for Real-Time Control and Neural Modeling. Chapter 13 in Handbook of Intelligent Control: Neural, Fuzzy and Adaptive Approaches (White & Sofge, eds.), Van Nostrand Reinhold, New York, NY, 1992.
48. Werbos, P., New Directions in ACDs: Key to Intelligent Control an Understanding the brain. Proceedings of the International Joint Conference on Neural Networks, Vol. 3, 2000, pp. 61–66.
49. Salcedo-Sanz, S., Santiago-Mozos, R., Bousoño-Calzón, C., A Hybrid Hopfield Network-Simulated Annealing Approach for Frequency Assignment in Satellite Communications Systems. IEEE Transactions on Systems, Man, and Cybernetics. Part B, Vol. 34, No. 2, April 2004, pp. 1108–1116.

50. Xiaowei Sheng, Minghu Jiang. A Information Retrieval System Based on Automatic Query Expansion and Hopfield Network. Proceedings of the IEEE International Neural Networks and Signal Processing, Nanjing, China, December 14–17, 2003, pp. 1624–1627.
52. Carpenter, G. A., and Grossberg, S., A Massively Parallel Architecture for a Self-Organizing Neural Pattern Recognition Machine. Computer Vision, Graphics and Image Processing, 37, 1987, pp. 54–115.
53. Chiung-Yao Fang, Sei-Wang Chen, Chiou-Shann Fuh. Automatic Change Detection of Driving Environments in a Vision-Based Driver Assistance System. IEEE Transactions on Neural Networks, Vol. 14, No. 3, May 2003, pp. 646–657.
54. Mulder, S. A., Wunsch, D. C., Using Adaptive Resonance Theory and Local Optimization to Divide and Conquer Large Scale Traveling Salesman Problems. Proceedings of the International Joint Conference on Neural Networks, IJCNN?03, 20–24 July 2003, Vol. 2, pp. 1408–1411.
55. Yu-Hu Cheng, Jian-Qiang Yi, Dong-Bin Zhao. Sensor-Based Navigation System Using a Modified AER-2 Network and Fuzzy Controller for an Indoor Mobile Robot. Proceedings of the Second International Conference on Machine Learning and Cybernetics, Xi'an, 2–5 November 2003, pp. 1374–1379.
56. Sugisaka, M., Dai, F., The Application of ART Neural Network to Image Processing for Controlling a Mobile Vehicle. Proceedings of the IEEE International Symposium on Industrial Electronics, ISIE 2001, 12–16 June 2001, Vol. 3, pp. 1951–1955.
57. Siew Chan Woo, Chee Peng Lim, R Osman. Development of a Speaker Recognition System using Wavelets and Artificial Neural Networks. Proceedings of the International Symposium on Intelligent Multimedia, Video and Speech Processing, Hong Kong, 2–4 May 2001, pp. 413–416.
58. Morra, L., Lamberti, F., Demartini, C., A Neural Network Approach to Unsupervised Segmentation of Single-Channel MR Images. Proceedings of the 1st International IEEE EMBS Conference on Neural Engineering, Capri Island, Italy, March 20–22, 2003, pp. 515–518.
59. Postolache, O., Girao, P., Pereira, S., Ramos, H., Wireless Water Quality Monitoring System Based on Field Point Technology and Kohonen Maps. Proceedings of IEEE Canadian Conference on Electrical and Computer Engineering, Montreal, May 4–7, Vol. 3, 2003, pp. 1872–1876.
60. Li Pan, Hong Zheng, Saeid Nahavandi. The Application of Rough Set and Kohonen Network to Feature Selection for Object Extraction. Proceedings of the Second In ternational Conference on Machine Learning and Cybernetics, Xi'an, 2–5 November 2003, pp. 1185–1189.
61. Sudha, N., An ASIC Implementation of Kohonen's Map based Color Image Compression. Proceedings of the 17th IEEE International Conference on VLSI Design (VLSID'04), 2004, pp. 4.
62. Hulsen, H., Garnica, St., Fatikow, S., Extended Kohonen Networks for the Pose Control of Microrobots in a Nanohandling Station. Proceedings of the 2003 IEEE International Symposium on Intelligent Control, Houston, Texas, Oct. 5-8, 2003, pp. 116–121.
63. Hubel, D. H., and Wiesel, T. N., Receptive Fields Binocular Interaction and Functional Architecture in the Cat's Visual Cortex. J. Physiology (Lond.), 1962, Vol. 106, No. 1, pp. 106–154.
64. Hubel, D. H., and Wiesel, T. N., Receptive Fields and Functional Architecture in Two Nonstriate Visual Areas (18 and 19) of the Cat. J. Neurophysiology, 1965, Vol. 28, No. 2, pp. 229–289.
65. Watanabe, A., Andoh, M., Chujo, N., Harata, Y., Neocognitron Capable of Position Detection and Vehicle Recognition. Proceedings of the International Joint Conference on Neural Networks, IJCNN?99, 10–16 July 1999, Vol. 5, pp. 3170–3173.
66. Fukushima, K., Neocognitron of a New Version: Handwritten Digit Recognition. Proceedings of the International Joint Conference on Neural Networks, IJCNN?01, 15–19 July 2001, Vol. 2, pp. 1498–1503.
67. San Kan Lee, Pau-choo Chung, Chein-I Chang, Chien-Shun Lo, Tain Lee, Giu-Cheng Hsu, Chin-Wen Yang. A Shape Cognitron Neural Network for Breast Cancer Detection.

Proceedings of the International Joint Conference on Neural Networks, IJCNN?02, 12–17 May 2002, Vol. 1 pp. 822–827.
68. Fukushima, K., Neural Network Models for Vision. Proceedings of the International Joint Conference on Neural Networks, IJCNN?03, 20–24 July 2003, pp. 2625–2630.
69. Bintu G. Vasudevan, Bhawani S. Gohil, Vijay K. Agarwal, Backpropagation Neural-Network-Based Retrieval of Atmospheric Water Vapor and Cloud Liquid Water From IRS-P4 MSMR, IEEE Transactions on Geoscience and Remote Sensing, Vol. 42, N 5, May 2004, pp. 985–990.
70. Hu Y. H. and Hwang J.-N., Eds., Handbook of Neural Network Signal Processing, Boca Raton, FL: CRC, 2002.
71. Betker, A. L., Moussavi, Z., Szturm, T., On Classification of Simulated Sensory Deficiency, Canadian Conference on Electrical and Computer Engineering, Vol. 2, 2–5 May 2004, pp. 1211–1214.
72. Helena Grip, Fredrik Ohberg, Urban Wiklund, Ylva Sterner, J. Stefan Karlsson, Bjorn Gerdle, Classification of Neck Movement Patterns Related to Whiplash-Associated Disorders Using Neural Networks, IEEE Transactions on Information Technology in Biomedicine, Vol. 7, No. 4, December 2003, pp. 412–418.
73. Jiebo Luo, Stephen P. Etz, A Physical Model-Based Approach to Detecting Sky in Photographic Images, IEEE Transactions on Image Processing, Vol. 11, No. 3, March 2002, pp. 201–212.
74. Dennis S.S. Lee, Brian J. Lightgow, Rob E.Morrison, New Fault Diagnosis of Circuit Breakers, IEEE Transactions on Power Delivery, Vol. 18, No. 2, April 2003, pp. 454–459.
75. Werbos, P., Neurocontrol and Supervised Learning: An Overview and Evaluation, Ed. by David A. White, Donald A. Sofge, In: Handbook on Intelligent Control: Neural, Fuzzy, and Adaptive Approaches, Part 3, 1992, Van Nostrand Reinhold, pp. 65–89.
76. Gaurav Saini, S.N. Balakrishnan, Adaptive Critic Based Neurocontroller for Autolanding of Aircrafts, Proceedings of the American Control Conference, Albuquerque, New Mexico, USA, June 1997, pp. 1081–1085.
77. Dongchen Han, S.N. Balakrishnan, State-Constrained Agile Missile Control with Adaptive-Critic-Based Neural networks, IEEE Transactions on Control Systems Technology, Vol. 10, No. 4, July 2002, pp. 481–489.
78. Venayagamoorthy, G. K., Harley, R. G., Wunsch, D. C., Implementation of Adaptive Critic-Based Neurocontrollers for Turbogenerators in a Multimachine Power System, IEEE Transactions on Neural Networks, Vol. 14, Issue 5, Sept. 2003, pp. 1047– 1064.
79. Xiaoqun Liao, Ernst L. Hall, Beyond Adaptive Critic - Creative Learning for Intelligent Autonomous Mobile Robots, Proceedings of ANNIE 2002, St. Louis, Missouri, USA, November 2002, pp. 1–6.
80. Kussul, E. M., Baidyk, T. N., Lukovitch, V. V., Rachkovskij, D. A., Adaptive high performance classifier based on random threshold neurons. Proc. of Twelfth European Meeting on Cybernetics and Systems Research (EMCSR-94), Austria, Vienna, April 5–8, 1994. In: R. Trappl, ed., Cybernetics and Systems 94, World Scientific Publishing Co. Pte. Ltd, Singapore, pp. 1687–1695.
81. Kussul, E. M., Baidyk, T. N., Neural Random Threshold Classifier in OCR Application, Proceedings of the Second All-Ukrainian Intern. Conf. UkrOBRAZ'94, Kyjiv, Ukraine, December 20–24, 1994, pp. 154–157.
82. Baidyk, T., Kussul, E., Makeyev, O., Texture Recognition with Random Subspace Neural Classifier, *WSEAS Transactions on Circuits and Systems*, Issue 4, Vol. 4, April 2005, pp. 319–325.
83. Baidyk, T., and Kussul, E., Neural Network Based Vision System for Micro Workpieces Manufacturing. *WSEAS Transactions on Systems*, Issue 2, Volume 3, April 2004, pp. 483–488.
84. Kussul, E., Baidyk, T., Wunsch, D., Makeyev, O., Martín, A., Permutation Coding Technique for Image Recognition Systems. *IEEE Transactions on Neural Networks*, Vol. 17/6, November 2006, pp. 1566–1579.

Chapter 3
Neural Classifiers

In this chapter we shall describe the neural classifiers. One of the important tasks in micromechanics for process automation is pattern recognition. For this purpose we developed different neural classifiers. Below, we will describe the Random Threshold Classifier (RTC classifier), Random Subspace Classifier (RSC classifier), and LIRA classifier (LImited Receptive Area). We will describe the structure and functions of these classifiers and how we use them. The first problem is the texture recognition problem.

3.1 RTC and RSC Neural Classifiers for Texture Recognition

The task of classification in recognition systems is a more important issue than clustering or unsupervised segmentation in a vast majority of applications [1]. Texture classification plays an important role in outdoor scene images recognition, surface visual inspection systems, and so on. Despite its potential importance, there is no formal definition of texture due to an infinite diversity of texture samples. There exists a large number of texture analysis methods in the literature.

On the basis of the texture classification, Castano et al. obtained satisfactory results for real-world image analysis relevant to navigation on cross-country terrain [1]. They had four classes: soil, trees, bushes/grass, and sky. This task was elected by Pietikäinen et al. to test their system for texture recognition [2]. In this case, a new database was created. Five texture classes were defined: sky, trees, grass, roads, and buildings. Due to perceptible changes of illumination, the following sub-classes were used: trees in the sun, grass in the sun, road in the sun, and buildings in the sun. They achieved a very good accuracy of 85.43%. In 1991, we solved a similar task [3, 4]. We worked with five textures (sky, trees/crown, road, transport means, and post/trunk). The images were taken in the streets of the city. We took brightness, contrast, and contour orientation histograms as input to our

DOI 10.1007/978-3-642-02535-8_3, © Springer-Verlag Berlin Heidelberg 2010

Fig. 3.1 Samples of real-world images

system (74 features). We used associative-projective neural networks for recognition [4], achieving a recognition rate of 79.9 %. In Fig. 3.1 two examples of our images are presented.

In 1996, A. Goltsev developed an assembly neural network for texture segmentation [5, 6] and used it for real scene analysis. Texture recognition algorithms are used in different areas, for example, in the textile industry for detection of fabric defects [7]. In the electronic industry, texture recognition is important to characterize the microstructure of metal films deposited on flat substrates [8], and to automate the visual inspection of magnetic disks for quality control [9]. Texture recognition is used for foreign object detection (for example, contaminants in food, such as pieces of stone, fragments of glass, etc.) [10]. Aerial texture classification is applied to resolve difficult figure-ground separation problems [11].

Different approaches were developed to solve the problem of texture recognition. Leung et al. [12] proposed textons (representative texture elements) for texture description and recognition. The vocabulary of textons corresponds to the characteristic features of the image. They tested this algorithm on the Columbia-Utrecht Reflectance and Textures (CUReT) database [13, 14]. This approach has disadvantages: it needs many parameters to be set manually and it is computationally complex. Another approach is based on the micro-textons, which are extracted by means of a multiresolution local binary pattern operator (LBP). LBP is a grayscale invariant primitive statistic of texture. This method was tested on CUReT database and performed well in both experiments and analysis of outdoor scene images.

Many statistical texture descriptors were based on a generation of co-occurrence matrices. In [9] the texture co-occurrence of *n*th order rank was proposed. This matrix contains statistics of the pixel under investigation and its surrounding pixels. The co-occurrence operator can be used to map the binary image, too. For example, in [15], the method to extract texture features in terms of the occurrence of *n* conjoint pixel values was combined with a single-layer neural network. There are many investigations in the application of neural networks for texture recognition [16, 17]. To test the developed system [17], texture images from [18] were used.

The reasons to choose a system based on neural network architecture include its significant properties of adaptiveness and robustness to texture variety.

3.1.1 *Random Threshold Neural Classifier*

The RTC neural classifier was developed and tested in 1994 [19]. The architecture of RTC is shown in Fig. 3.2.

The neural network structure consists of s blocks, and each block has one output neuron ($b_1, \ldots, b_s$). The set of features ($X_1, \ldots, X_n$) is input to every block. Every feature X_i is input to two neurons h_i^j and l_i^j, where i ($i = 1, 2, \ldots, n$) represents the number of features, and j ($j = 1, 2, \ldots, s$) represents the number of neural blocks. The threshold of l_i^j is less than the threshold of h_i^j. The values of thresholds are randomly selected once and fixed. The output of neuron l_i^j is connected with the excitatory input of the neuron a_i^j, and the output of the neuron h_i^j is connected with the inhibitory input of the neuron a_i^j. In the output of the neuron a_i^j, the signal appears only if the input signal from l_i^j is equal to 1 and the input signal from h_i^j is equal to 0. All outputs from neurons a_i^j, in one block j, are inputs of the neuron b_j, which presents the output of the whole neuron block.

The output of the neuron b_j is 1 only if all neurons a_i^j in the block j are excited. The output of every neuron block is connected with trainable connections to all the inputs of the output layer of the classifier ($c_1, \ldots, c_t$), where t is a number of classes.

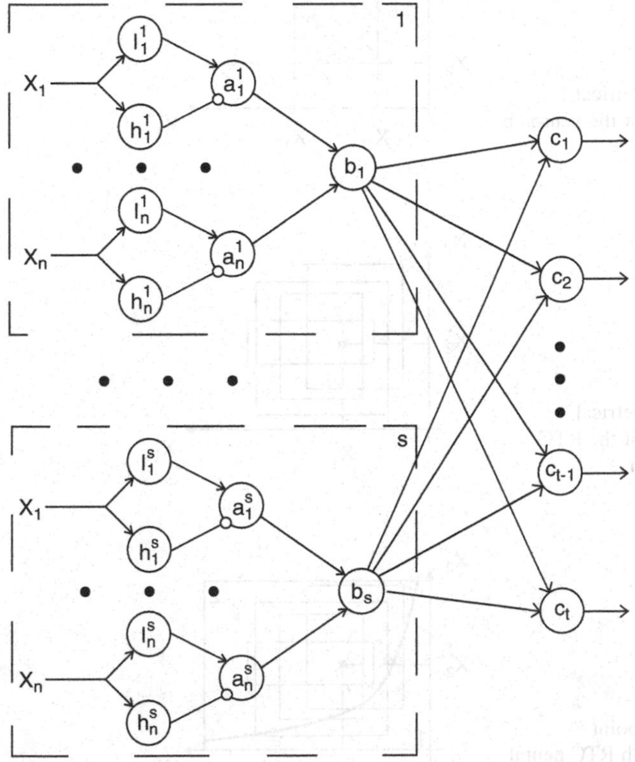

Fig. 3.2 Structure of RTC neural classifier

The classifier works in two modes: training and recognition. We use the perceptron rule to change the connection weights during the training process.

The geometrical interpretation can help us explain the discussed principles. Let us consider the case with two features X_1 and X_2 (Fig. 3.3). The neuron b_1 is active when the point representing input features is located inside the rectangle shown in Fig. 3.3. Since there are many blocks in RTC (1, ..., s), the whole feature space will be covered by many rectangles of different sizes and locations (Fig. 3.4). In multidimensional space, instead of rectangles we will get multidimensional parallelepipeds. Each parallelepiped corresponds to the active state of one b_j neuron (Fig. 3.2).

For example, if we want to recognize the new point (X_1^*, X_2^*) in the space of two features (X_1 and X_2) containing two classes (1 and 2) (Fig. 3.5), we will obtain the neuron responses which are related to the rectangles that cover the new point. During the training process, connections of active neurons that correspond to point (X_1^*, X_2^*) with output neurons that correspond to the second class will be stronger than those with output neurons that correspond to the first class because a major part

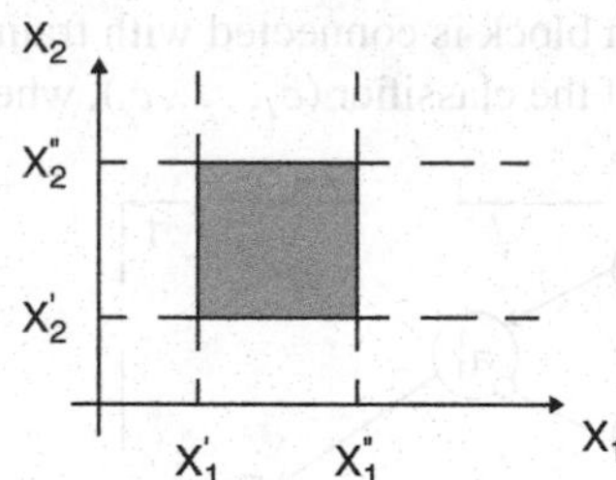

Fig. 3.3 Geometrical interpretation of the neuron b activity

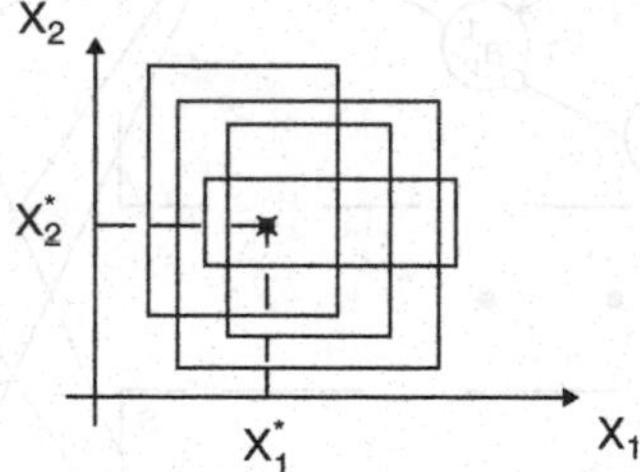

Fig. 3.4 Geometrical interpretation of the RTC neural classifier

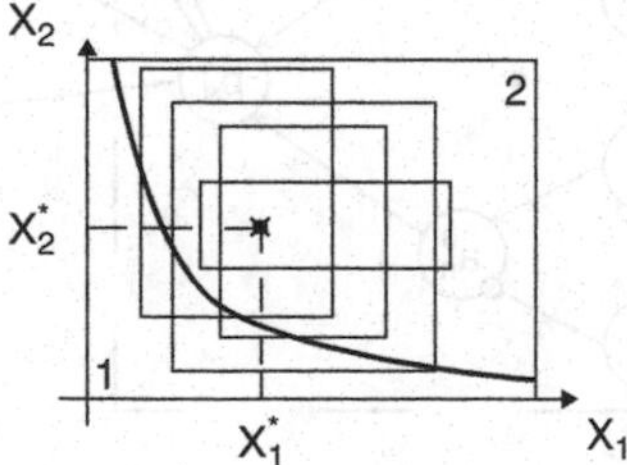

Fig. 3.5 New point recognition with RTC neural classifier

of these rectangles is covered by the second class area. Therefore, this point will be recognized as a point of the second class.

3.1.2 Random Subspace Classifier

When the dimension of input space n (Fig. 3.2) increases, it is necessary to increase the gap between the thresholds of neurons h_i^j and l_i^j, so for large n, many thresholds of neurons h_i^j achieve the higher limit of variable X_i, and many thresholds of l_i^j achieve the lower limit of variable X_i. In this case, the corresponding neuron a_i^j always has output 1 and gives no information about the input data. Only a small part of neurons a_i^j can change the outputs. To save calculation time, we have modified RTC classifiers including for each block from 1 to S not all the input variables $(X_1, \ldots, X_n)$, but only a small part of them, selected randomly. This small number of chosen components from the input vector we term the random subspace of the input space. For each block j we select different random subspaces. Thus, we represent our input space by a multitude of random subspaces.

The Random Subspace Classifier was developed for the general classification problem in parameter spaces of limited dimensions. The structure of this classifier is presented in Fig. 3.6.

This classifier contains four neural layers: the input layer $X = x_1, x_2, \ldots, x_K$; the intermediate layer $GROUP = group_1, group_2, \ldots, group_N$; the associative layer $A = a_1, a_2, \ldots, a_N$; the output layer $Y = y_1, y_2, \ldots, y_m$. Each neuron x_i of the input layer corresponds to the component of the input vector to be classified. Each $group_i$ of the intermediate layer contains some quantity P of neuron pairs p_{ij}. Each pair p_{ij} contains one ON neuron and one OFF neuron (Fig. 3.6). The ON neuron is active if $x_r > T_{ONij}$. Each OFF neuron is active if $x_r < T_{OFFij}$, where T_{OFFij} is the threshold of the OFF-neuron and T_{ONij} is the threshold of the ON neuron. Each pair p_{ij} is connected with a randomly selected neuron of the input layer X. All neuron thresholds of the layer $GROUP$ are selected randomly under condition $T_{ONij} < T_{OFFij}$ in each pair. All the neurons of $group_i$ are connected with one neuron a_i of the associative layer A. A neuron a_i is active if and only if all the neurons of $group_i$ are active. The output of the active neuron equals 1. If the neuron is not active, its output equals 0. Each neuron of the A layer is connected with all neurons of the output layer Y. The training process changes the weights of these connections. The training and the winner selection rules are the same as in the classifier LIRA.

The main difference between the classifier LIRA and the RSC classifier is the absence of the group layer. Instead of each pair of the neurons in the layer $GROUP$, the classifier LIRA uses one connection, either ON type or OFF type, which could be active or inactive. This modification permits an increase in the classification speed. Another difference is related to the applications of these classifiers. We applied the RSC classifier for the texture recognition and other problems where the activities of the input neurons were calculated with a special algorithm of the feature extraction. The classifier LIRA is applied directly to a raw image.

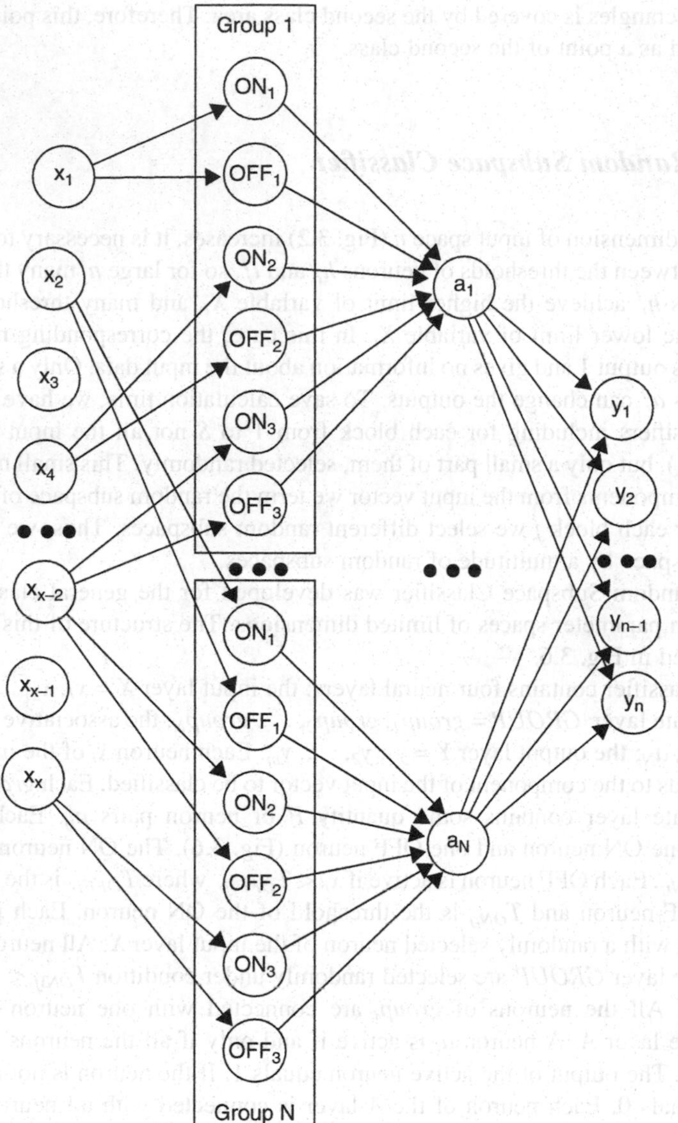

Fig. 3.6 The structure of the Random Subspace Classifier

3.1.3 *Encoder of Features*

The task of an encoder is to codify the feature vector $(X_1, \ldots, X_n)$ into binary form in order to present it to the input of the one-layer classifier. The feature encoder calculates the output values of the neurons b1, …, bs.

To create the encoder structure, we have to select the subspaces for each neuron block. For example, if the subspace size is 3, in each neuron block j we will use only three input parameters whose numbers we select randomly from the range 1, . . ., n (where n is the dimension of the input space; in our case, n = 48). After that, we calculate the thresholds for each pair of neurons l_i^j and h_i^j of three selected neurons a_i^j of the block j. For this purpose, we select the point x_i^j randomly from the range [0, . . ., X_i]. After that, we select the random number y_i^j uniformly distributed in the range [0, . . ., GAP], where GAP is the parameter of the encoder structure. Then we calculate the thresholds of neurons l_i^j and h_i^j in accordance with the following formulas:

$$\begin{aligned} &Trl_i^j = x_i^j - y_i^j; \\ &if\left(Trl_i^j < X_i \min\right) \; then \; Trl_i^j = X_i \min; \end{aligned} \tag{3.1}$$

$$\begin{aligned} &Trh_i^j = x_i^j + y_i^j; \\ &if\left(Trh_i^j > X_i \max\right) \; then \; Trh_i^j = X_i \max; \end{aligned} \tag{3.2}$$

where Trl_i^j and Trh_i^j are the thresholds of neurons l_i^j and h_i^j respectively, and $X_i min$ and $X_i max$ are the minimum and maximum possible values for a component X_i of the input vector (X_1, . . ., X_n). Then the encoder forms a binary vector (b_1, . . ., b_s) corresponding to the feature vector. This vector is presented to the input of the one-layer classifier. The training rule of our one-layer classifier is the same as the training rule of the one-layer perceptron.

3.2 LIRA Neural Classifier for Handwritten Digit Recognition

There are many applications for handwritten digit recognition, such as for bank checks and custom declaration automatic reading. Various methods were proposed to solve this problem [20–22]. To estimate a method's effectiveness, the most important parameter is recognition rate. This parameter shows the proportion of samples in the test database that is recognized correctly.

The MNIST database contains 60,000 handwritten digits in the training set and 10,000 handwritten digits in the test set. Different classifiers tested on this database by LeCun [21] had shown recognition rates from 88% to 99.3% (Table 3.1). In recent years, new results were obtained using shape matching and Support Vector Machine methods [23, 24].

We have developed the new neural classifier LIRA (LImited Receptive Area classifier) based on Rosenblatt's perceptron principles. To adapt Rosenblatt's perceptron for handwritten digit recognition problems, we made some changes in perceptron structure, training, and recognition algorithms. Rosenblatt's perceptron contains three layers of neurons. The first layer S corresponds to the retina. In technical terms it corresponds to input image. The second layer A, called the

Table 3.1. Error rates of different classifiers on the MNIST database

Methods	% of error number	Ref.	Year
K Nearest Neighbor	2.4	[25]	1994
Fully connected net	1.6	[25]	1994
LeNet-1	1.7	[25]	1994
LeNet-4	1.1	[25]	1994
LeNet-4 / Local Learning	1.1	[25]	1994
Reduced Set SVM poly 5	1.0	[21]	1998
LeNet-5	0.95	[21]	1998
Virtual SVM poly 9 [distortions]	0.8	[21]	1998
LeNet-5 [distortions]	0.8	[21]	1998
SVC-rbf_binary	0.72	[21]	1998
Boosted LeNet-4 [distortions]	0.7	[25]	1994
Shape Matching + 3-NN	0.63	[23],[24]	2001
Classifier LIRA	0.63	[26]	2002
Classifier LIRA	**0.55**	[27]	2004
Classifier with permutation coding	**0.44**	[28]	2006
SVC-rbf_grayscale	**0.42**	[29]	2002

associative layer, corresponds to the feature extraction subsystem. The third layer *R* corresponds to the output of the entire system. Each neuron of this layer corresponds to one of the output classes. In handwritten digit recognition tasks, this layer contains 10 neurons corresponding to digits 0 through 9. Connections between the layers *S* and *A* are established using a random procedure and cannot be changed by perceptron training. They have the weights 0 or 1.

Each neuron of the *A*-layer is connected with all the neurons of the *R*-layer. Initially the weights are set to 0, but they are changed during the perceptron training. The rule of weights changing corresponds to the training algorithm. We used a training algorithm slightly different from Rosenblatt's. We have also modified the random procedure of *S*-connection establishment. Our latest modifications are related to the rule of winner selection in the output *R*-layer. In this section, we describe our approach and the handwritten digit recognition results.

3.2.1 Rosenblatt Perceptrons

3-layer Rosenblatt perceptrons contain a sensor layer *S*, an associative layer *A*, and a reaction layer *R*. Many investigations were dedicated to perceptrons with one neuron in layer *R* (*R*-layer) [30]. Such perceptrons can recognize only two classes. If the output of an *R* neuron is higher than the predetermined threshold *T*, the input image belongs to class 1. If it is lower than *T*, the input image belongs to class 2. The sensor layer *S* (*S*-layer) contains two $\{-1, 1\}$ state elements. The element is set to 1 if it belongs to the object image and set to –1 if it belongs to the background.

The associative layer *A* (*A*-layer) contains neurons with two $\{0, 1\}$ state outputs. Inputs of these neurons are connected with outputs of *S*-layer neurons with no modifiable connections. Each connection may have the weight 1 (positive connection) or the weight –1 (negative connection). Let the threshold of such neurons equal the number of its input connections. This neuron is active only if all positive connections correspond to the object and all negative connections correspond to the background.

The neuron R is connected with all neurons of A-layer. The weights of these connections are changed during the perceptron training. The most popular training rule is increasing the weights between active neurons of A-layer and neuron R if the object belongs to class 1. If the object belongs to class 2, the corresponding weights are decreasing. It is known that such perceptrons have fast convergences and can form nonlinear discriminating surfaces. The complexity of discriminating surfaces depends on the number of A-layer neurons.

3.2.2 Description of the Rosenblatt Perceptron Modifications

We have proposed several changes to perceptron structure to create the neural classifiers for handwritten digit recognition. To examine them we used an MNIST database [21]. Each black-and-white digit image is presented by a 20*20 pixel box. The image was converted to the gray level and was centered in a 28*28 image by computing the center of mass of the pixels and translating the image so as to position this point at the center of the 28*28 field. In our case, we worked with the binary image.

A binary image is obtained from the gray-level image using the following procedure. The threshold th is computed as:

$$th = \frac{2 * (\sum_{i=1}^{W_S} \sum_{j=1}^{H_S} b_{ij})}{W_S \cdot H_S} \tag{3.3}$$

where H_S is the number of rows of the image; W_S is the number of columns of the image; b_{ij} is the brightness of the pixel of the grayscale image; and s_{ij} is the brightness of the pixel of the resulting binary image:

$$s_{ij} = \begin{cases} 1, if\ b_{ij} > th, \\ -1, if\ b_{ij} \leq th. \end{cases} \tag{3.4}$$

For the MNIST database, $H_S = W_S = 28$.

For the first modification of a simple Rosenblatt perceptron, ten neurons were included in R-layer. In this case, it is necessary to introduce the rule of winner selection. In the first series of experiments, we used the simplest rule of winner selection: the neuron from R-layer having the highest excitation determines the class under recognition. Using this rule, we obtained a recognition rate of 99.21%. After that, we modified the winner selection rule and improved the recognition rate to 99.37%. We will describe this selection rule later.

The second modification was made in the training process. Let the neuron-winner have excitation E_w and its nearest competitor have excitation E_c. If

$$(\mathrm{E_w} - E_c)/E_w < T_E, \tag{3.5}$$

the competitor is considered a winner, where T_E is the superfluous excitation of the neuron-winner.

The third modification is concerned with connections. The connections between A-layer and R-layer of a Rosenblatt perceptron can be negative or positive. We used only positive connections. In this case, the training procedure is the following:

1. Let j correspond to the correct class under recognition. During the recognition process we obtain excitations of R-layer neurons. The excitation of neuron R_j corresponding to the correct class is decreased by the factor (1 - T_E). After this the neuron having maximum excitation R_k is selected as winner.
2. If $j = k$, nothing to be done.
3. If j does not equal k,

$$w_{ij}(t+\ 1) = w_{ij}(t) + a_i, \tag{3.6}$$

where $w_{ij}(t)$ is the weight of the connection between the i neuron of the A-layer and the j neuron of the R-layer before modification, $w_{ij}(t+1)$ is the weight after modification, and a_i is the output signal (0 or 1) of the i neuron of the A-layer.

$$\begin{aligned} w_{ik}(t+\ 1) = w_{ik}(t) - a_i, \qquad & \text{if } (w_{ik}(t)0), \\ w_{ik}(t+\ 1) = 0, \qquad & \text{if } (w_{ik}(t) = 0), \end{aligned} \tag{3.7}$$

where $w_{ik}(t)$ is the weight of the connection between the i neuron of the A-layer and the k neuron of the R-layer before modification, and $w_{ik}(t+1)$ is the weight after modification. A more detailed description of the training procedure will be offered later. The perceptron with these changes is termed the LImited Receptive Area classifier for binary images (LIRA_binary) (Fig. 3.7). A more general case of such a classifier was developed and named Random Subspace Classifier (RSC) [19, 31–33].

Each A-layer neuron of LIRA has random connections with the S-layer. To install these connections, it is necessary to enumerate all elements of the S-layer. Let the number of these elements equal N_S. To determine the connection of the A-layer neuron, we select a random number uniformly distributed in the range $[1, N_S]$. This number determines the S-layer neuron, which will be connected with the mentioned A-layer neuron. The same rule, which was proposed by Rosenblatt [30], is used to determine all connections between A-layer neurons and S-layer neurons. Our experience shows that it is possible to improve the perceptron performance by modifying this rule.

The fourth modification is the following: We connect A-layer neurons with S-layer neurons randomly selected not from the entire S-layer, but rather from the

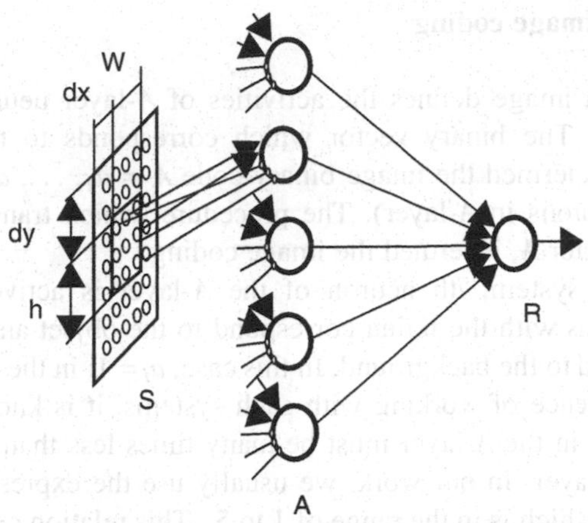

Fig. 3.7 LImited Receptive Area (LIRA) classifier

rectangle ($h * w$), which is located in the S-layer (Fig. 3.7). The distances dx and dy are random numbers selected from the ranges: dx from $[0, W_S - w]$ and dy from $[0, H_S - h]$, where W_S and H_S stand for the width and height of S-layer.

3.2.2.1 Mask Design

The associative neuron mask is the number of positive and negative connections of an A-layer neuron with the retina. The procedure of random selection of connections is used to design the mask. This procedure begins from choice of the upper left corner of the rectangle in which all positive and negative connections of the associative neuron are located. These formulas are used:

$$dx_i = random_i(W_S - w),$$
$$dy_i = random_i(H_S - h), \tag{3.8}$$

where i is the position of a neuron in associative layer A and $random_i(z)$ is the random number which is uniformly distributed in the range $[0, z]$. After that, each positive and negative connection position within the rectangle is defined by a couple of numbers:

$$x_{ij} = random_{ij}(w),$$
$$y_{ij} = random_{ij}(h), \tag{3.9}$$

where j is the number of ith neuron connection with the retina. Absolute coordinates of the connection on the retina are defined by a couple of the numbers:

$$X_{ij} = x_{ij} + d_{xi},$$
$$Y_{ij} = y_{ij} + dy_i. \tag{3.10}$$

3.2.2.2 Image coding

Any input image defines the activities of A-layer neurons in one-to-one correspondence. The binary vector which corresponds to the activity of associative neurons is termed the image binary code $A = a_1, \ldots, a_n$, (where n is the number of the neurons in A-layer). The procedure, which transforms the input image to binary vector A, is termed the image coding.

In our system, ith neuron of the A-layer is active only if all the positive connections with the retina correspond to the object and all negative connections correspond to the background. In this case, $a_i = 1$; in the opposite case, $a_i = 0$. From the experience of working with such systems, it is known that the active neuron number m in the A-layer must be many times less than the whole neuron number n of this layer. In our work, we usually use the expression $m = c\sqrt{n}$, where c is constant, which is in the range of 1 to 5. This relation corresponds to neurophysiological facts. The number of active neurons in the cerebral cortex is hundreds of times less than the total number of neurons.

Taking into account the small number of active neurons, it is convenient to represent the binary vector $\boldsymbol{A}$ not explicitly but as a list of numbers of active neurons. For example, let the vector A equal the following:

$$\boldsymbol{A} = 00010000100000010000.$$

The corresponding list of the numbers of active neurons will be 4, 9, and 16. This list is used for saving the image codes in compact form and for the fast calculation of activity of the neurons of the output layer. Thus, after execution of the coding procedure, every image has a corresponding list of numbers of active neurons.

3.2.2.3 Training procedure

Before training, all the weights of connections between neurons of the A-layer and the R-layer are set to 0.

1. The training procedure begins with the presentation of the first image to the perceptron. The image is coded and the R-layer neuron excitation E_i is computed. E_i is defined as:

$$E_i = \sum_{j=1}^{n} a_j * w_{ji}, \tag{3.11}$$

where E_i is the excitation of the ith neuron of the R-layer; a_j is the excitation of the jth neuron of the A-layer; and w_{ji} is the weight of the connection between the jth neuron of the A-layer and the ith neuron of the R-layer.

2. We require recognition to be robust. After the calculation of excitations of all R-layer neurons, the correct name of the presented image is read from the mark

file of the MNIST database. The excitation E of the corresponding neuron is recalculated according to the formula:

$$E_k^* = E_k * (1 - T_E). \tag{3.12}$$

After that, we find the neuron (winner) with the maximal activity. This neuron represents the recognized handwritten digit.

3. We denote the neuron-winner number as i_w, and the number of the neuron, which really corresponds to the input image, as i_c. If $i_w = i_c$, nothing is done. If $i_w \neq i_c$,

$$\begin{aligned} &(\forall j)\ \left(w_{ji_c}(t+1) = w_{ji_c}(t) + a_j\right) \\ &(\forall j)\ \left(w_{ji_w}(t+1) = w_{ji_w}(t) - a_j\right) \\ &\text{if}(w_{ji_w}(t+1) < 0)\ w_{ij_w}(t+1) = 0, \end{aligned} \tag{3.13}$$

where $w_{ji}(t)$ is the weight of the connection between the j neuron of the A-layer and the i neuron of the R-layer before reinforcement, and $w_{ji}(t + 1)$ is the weight after reinforcement.

The training process is carried out iteratively. After representation of all the images from the training subset, the total number of training errors is calculated. If this number is higher than 1% of the total number of images, then the next training cycle is executed. If the error number is less than 1%, the training process is stopped. The training process is also stopped when the cycle number is more than the previously prescribed value. In previous experiments this value was 10 cycles, and in later ones it was 40 cycles.

It is obvious that in every new training cycle the image coding procedure is repeated and gives the same results as in previous cycles. Therefore, in final experiments we performed the coding process of images only once and recorded the lists of active neuron numbers for each image on a hard drive. Later, for all cycles we used not the images but corresponding lists of active neurons. Due to this procedure, the training process was accelerated approximately to an order of magnitude.

It is known [21] that the recognition rate of handwritten symbols may be increased significantly if the images are represented during the training cycle not only in their initial state but also by shifting and changing the image inclination (so called distortions). In our final experiments we used, besides the initial images, 16 variants of each image with distortions. Distortion models can be used to increase the effective size of a data set without actually collecting more data. We used 16 distortion variants (Fig. 3.8.): 12 shifts and four skewings. These skewing angles were selected: –26°, –13°, 13°, and 26°.

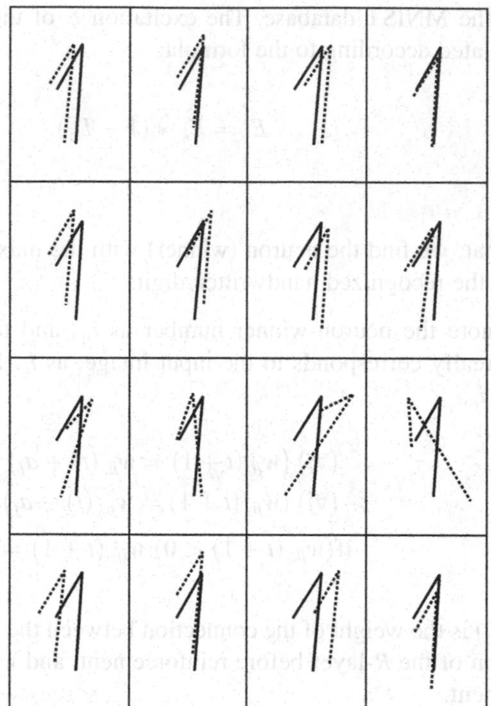

Fig. 3.8 The scheme of 16 distortions

3.2.2.4 Recognition procedure

To examine the recognition rate, the test set of the MNIST database was used. This test set contains 10,000 images. Coding and calculation of neuron activity were made by the same rules as by training, but the value T_E (reserve of robustness) was 0.

The recognition process for the new classifier differs from the previous ones. In this version, we use distortions in the recognition process, too. There is a difference between implementation of distortions during the training session and during the recognition session. During the training session, each new position of the initial image produced by distortions is considered a new image, which is independent of other image distortions. During the recognition session, it is necessary to introduce a rule for decision making. All the recognition results of one image and its distortions must be used for receiving one result, which gives the class name of the image under recognition. We have developed two rules for decision making.

Rule 1. According to this rule all the excitations of the *R*-layer neurons are summed for all the distortions.

$$E_i = \sum_{k=1}^{d} \sum_{j=1}^{n} a_{kj} * w_{ji} \tag{3.14}$$

where E_i is the excitation of the ith neuron of the R-layer; a_{kj} is the excitation of the jth neuron of the A-layer in the kth distortion; and w_{ji} is the weight of the connection between the jth neuron of the A-layer and the ith neuron of the R-layer. After that, the neuron-winner is selected as a result of recognition.

Rule 2. The second rule consists of calculations of R-layer neuron excitations and selection of the neuron-winner and its nearest competitor for each distortion. For the kth distortion, the relation r_k of the neuron-winner excitation E_{wk} to its nearest competitor excitation E_{ck} is calculated:

$$r_k = \frac{E_{wk}}{E_{ck}}. \tag{3.15}$$

After that, we select distortion with the maximal r_k. The neuron-winner of this distortion is considered to be the result of recognition.

3.3 LIRA-Grayscale Neural Classifier

To adapt the LIRA classifier for grayscale image recognition we have added additional neuron layer between the S-layer and the A-layer. We term it the I-layer (intermediate layer, see Fig. 3.9).

Each input of the I-layer neuron has one connection with the S-layer. Each output of this neuron is connected with the input of one neuron of the A-layer.

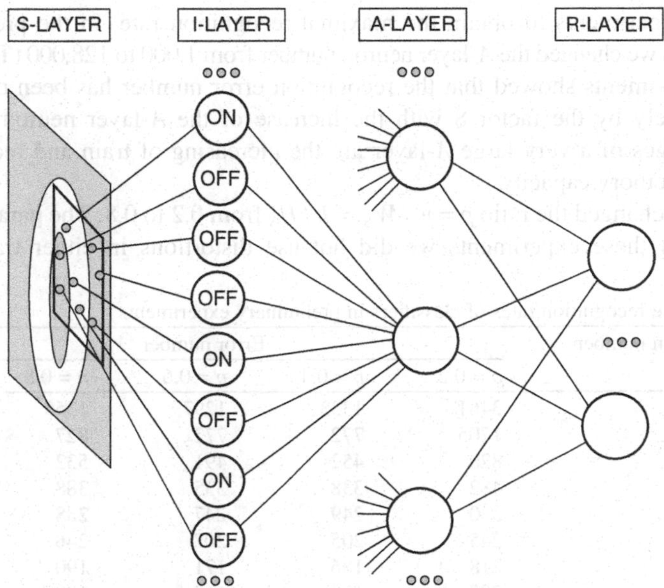

Fig. 3.9. LIRA-grayscale scheme

All the *I*-layer neurons connected to one *A*-layer neuron form the group of this *A*-layer neuron. The number of neurons in one group corresponds to the number of positive and negative connections between one neuron of the *A*-layer and the retina in the LIRA_binary structure. The *I*-layer neurons could be ON neurons or OFF neurons. The output of the ON neurons *i* is "1" when its input is higher than the threshold θ_i, and otherwise it equals "0." The OFF neuron *j* output is "1" when its input is less than the threshold θ_j, and otherwise it equals "0."

The ON neuron number in each group corresponds to the number of the positive connections of one *A*-layer neuron in the LIRA_binary structure. The OFF neuron number in each group corresponds to the number of the negative connections. In our case we selected three ON neurons and five OFF neurons. The rule of connection arrangement between the retina and one group of the *I*-layer is the same as the rule of mask design for one *A*-layer neuron in the LIRA-binary.

The thresholds θ_i and θ_j are selected randomly from the range $[0, \eta \cdot b_{\max}]$, where $b_{\max}$ is maximal brightness of the image pixels, and η is the parameter from [0, 1], which is selected experimentally. The output of the *A*-layer neuron is "1" if all outputs of its *I*-layer group are "1." If any neuron of this group has the output "0," the *A*-layer neuron has the output "0."

3.4 Handwritten Digit Recognition Results for Lira-binary

We carried out preliminary experiments to estimate the performance of our classifiers. On the basis of these experiments, we selected the best classifiers and carried out final experiments to obtain the maximal recognition rate. In the preliminary experiments we changed the *A*-layer neuron number from 1,000 to 128,000 (Table 3.2). These experiments showed that the recognition error number has been decreased approximately by the factor 8 with the increase of the *A*-layer neuron number. Disadvantages of a very large *A*-layer are the increasing of train and recognition time and memory capacity.

We also changed the ratio $p = w / W_S = h / H_S$ from 0.2 to 0.8. The parameter T_E was 0.1. In these experiments we did not use distortions in either training or

Table 3.2 The recognition rates of classifiers in preliminary experiments

A-layer neuron number	Error number				
	$p = 0.2$	$p = 0.4$	$p = 0.6$	$p = 0.8$	$p = 1$
1000	3461	1333	1297	1355	1864
2000	1705	772	772	827	1027
4000	828	452	491	532	622
8000	482	338	335	388	451
16000	330	249	247	288	337
32000	245	205	207	246	270
64000	218	186	171	190	217
128000	207	170	168	190	195

Table 3.3 Error numbers in the final experiments

Number of experiments	Without distortions	Rule 1		Rule 2	
		4 dist.	8 dist.	4 dist.	8 dist.
1	72	68	63	75	75
2	86	65	65	72	67
3	83	66	62	68	72
Mean value	80	66	**63**	72	71

recognition sessions. For each set of parameters, we made ten training cycles on the MNIST training set. After that, we estimated the recognition rate on the MNIST test set. The recognition rates obtained in the preliminary experiments are presented in Table 3.2.

In the preliminary experiments, we created three positive and three negative connections for each A-layer neuron. In the final experiments, we created three positive and five negative connections (Table 3.3). The number of A-layer neurons was 256,000. The window parameters were $w = 10$ and $h = 10$ and the retina size was 28 x 28. The number of training cycles was 40.

The coding time was 20 hours and the training time was 45 hours. The recognition time (for 10,000 samples) was 30 minutes without distortions, 60 minutes for four distortions, and 120 minutes for eight distortions. We conducted different experiments with different numbers of distortions during the recognition session (four and eight). We created distortions only with shifting (the first four or eight cases in Fig. 3.8). For comparison, we conducted experiments without distortions in the recognition session.

For statistical comparison we conducted three experiments for each set of parameters. The difference between the experiments consists of using different random structures of connections between the S-layer and the A-layer. In Table 3.3, each line corresponds to one of such experiments. The column title "Rule 1" corresponds to the first rule of winner selection (Formula 3.14) and "Rule 2" corresponds to the second rule of winner selection (Formula 3.15). The error number 63 corresponds to 99.37% of the recognition rate. Serge Belongie et al. [23] have the same result on the MNIST database.

3.5 Handwritten Digit Recognition Results for LIRA-Grayscale

The recognition rates obtained in the experiments with the LIRA-grayscale are presented in Table 3.4. In this case, we also conducted three experiments for statistical comparison. The difference between the experiments consists of using the different random structures of connections between the S-layer and the A-layer. In Table 3.4, each column is marked by the number of the experiment. The minimum number of errors we obtained was 59. This result is better than the result obtained with LIRA-binary.

Table 3.4 Error numbers in the final experiments

Number of experiments	1	2	3
Number of errors	60	59	64
Mean number of errors	61.33		

3.6 Discussion

The novel neural classifier LIRA was developed, which contains three neuron layers: sensor, associative, and output layers. The sensor layer is connected with the associative layer with no modifiable random connections, and the associative layer is connected with the output layer with trainable connections. The training process converges sufficiently fast. This classifier does not use floating-point and multiplication operations, and this property, in combination with the parallel structure of the classifier, permits it to be implemented in low-cost, high-speed electronic devices. The classifier LIRA, tested on the MNIST database, shows good recognition rates. It contains 60,000 handwritten digit images for the classifier training and 10,000 handwritten digit images for the classifier testing.

The results obtained on the MNIST database seem sufficient for applications, but there are many uses for handwritten number recognition. If the number contains, for example, ten digits and the recognition rate of one digit is 0.994 (as in our case), the whole number recognition rate could be $0.994^{10} = 0.942 = 94.2\%$. This recognition rate is insufficient for many applications. For this reason, additional investigations are needed to improve handwritten digit recognition rate.

The recognition time of each handwritten digit is also an important parameter. In many cases, to estimate the recognition time, the authors of different methods give the number of multiply-accumulate operations for one symbol recognition. For example, for the RS-SVM method it equals 650,000, and LeNet-5 is about 60% less expensive [21]. It is difficult to compare our classifier using this parameter because our classifier uses neither multiply operations nor floating-point operations. For one digit recognition, our classifier needs approximately 50,000 fixed-point add operations. This may seem very fast, but it is not. For one image coding, it needs approximately 10 x 256,000 readings from memory and logical operations. During recognition, we must code not only the initial image, but also, for example, four distortions. This whole process demands nearly ten million operations for each digit recognition, which is difficult to compare with the number of floating-point operations. In general, our classifier has lower recognition speed than methods by LeCun and SVM.

The other method to compare recognition time is testing classifiers on similar computers. Belongie [23] gives the time of the shape matching as 200 ms on a Pentium III, 500 MHz workstation. Using the regular nearest neighbor method, it is necessary to make N matchings for each digit recognition, where N is the size of the training set (Belongie used from 15,000 to 20,000 images). In this case, the recognition time of one digit could be from 3,000 seconds to 4,000 seconds.

We also tested our latest version of the classifier on a Pentium III, 500 MHz and obtained the recognition time of one digit in 0.5 seconds.

The third important parameter for classifier comparison is the training time. Belongie used the nearest neighbor classifiers, which practically need no training time [23]. The training time of LeNet was two to three days of CPU to train LeNet5 on a Silicon Graphics Origin 2000 server using a single R10000 200 MHz on processor. Our training time is 55 hours on a Pentium III, 500 MHz computer.

References

1. Castano R., Manduchi R., Fox J., Classification Experiments on Real-world Texture, Proceedings of the Third Workshop on Empirical Evaluation Methods in Computer Vision, Kauai, Hawaii, 2001, pp. 3–20.
2. Matti Pietikäinen, Tomi Nurmela, Topi Mäenpää, Markus Turtinen, View-based Recognition of Real-world Textures, *Pattern Recognition*, Vol.37, 2004, pp. 313–323.
3. Baidyk T., Comparison of Texture Recognition with Potential Functions and Neural Networks, *Neural Networks and Neurocomputers*, 1991, pp. 52–61 (in Russian).
4. Kussul E., Rachkovskij D., Baidyk T., On Image Texture Recognition by Associative-Projective Neurocomputer, Proceedings of the International Conference on Intelligent Engineering Systems through Artificial Neural Networks, 1991, pp. 453–458.
5. Goltsev A., An Assembly Neural Network for Texture Segmentation, *Neural Networks*, Vol. 9, No. 4, 1996, pp. 643–653.
6. Goltsev A., Wunsch D., Inhibitory Connections in the Assembly Neural Network for Texture Segmentation, *Neural Networks*, Vol. 11, 1998, pp. 951–962.
7. Chi-ho Chan, Grantham K.H. Pang, Fabric Defect Detection by Fourier Analysis, *IEEE Transactions on Industry Applications*, Vol. 36/5, 2000, pp. 1267–1276.
8. Zschech E., Besser P.. Microstructure Characterization of Metal Interconnects and Barrier Layers: Status and Future. Proc. of the IEEE International Interconnect Technology Conference, 2000, pp. 233–235.
9. Hepplewhite L., Stonham T.J., Surface Inspection Using Texture Recognition, Proc. of the 12th IAPR International Conference on Pattern Recognition, Vol. 1, 1994, pp. 589–591.
10. Patel D., Hannah I., Davies E.R., Foreign Object Detection via Texture Analysis, Proc. of the 12th IAPR International Conference on Pattern Recognition, 1994, Vol. 1, pp. 586–588.
11. Hoogs A., Collins R., Kaicic R., Mundy J., A Common Set of Perceptual Observables for Grouping, Figure-ground Discrimination, and Texture Classification, *IEEE Transactions on Pattern Analysis and Machine Intelligence*, Vol. 25/4, 2003, pp. 458–474.
12. Leung T., Malik J., Representing and Recognizing the Visual Appearance of Materials Using Three-dimensional Textons, *Int. J. Comput. Vision*, Vol. 43 (1), 2001, pp. 29–44.
13. Dana K.J., B. van Ginneken, Nayar S.K., Koendrink J.J., Reflectance and Texture of Real World Surfaces, *ACM Trans. Graphics*, Vol. 18 (1), 1999, pp. 1–34.
14. http://ww1.cs.columbia.edu/CAVE/curet/
15. Patel D., Stonham T.J., A Single Layer Neural Network for Texture Discrimination, IEEE International Symposium on Circuits and Systems, Vol. 5, 1991, pp. 2656–2660.
16. Mayorga M.A., Ludeman L.C., Shift and Rotation Invariant Texture Recognition with Neural Nets, Proc. of the IEEE International Conference on Neural Networks, Vol. 6, 1994, pp. 4078–4083.
17. Woobeom Lee, Wookhyun Kim, Self-organization Neural Networks for Multiple Texture Image Segmentation, Proc. of the IEEE Region 10 Conference TENCON 99, Vol. 1, 1999, pp. 730–733.

18. Brodatz P., Texture: A Photographic Album for Artists and Designers, Dover Publications, New York, 1966.
19. Kussul E., Baidyk T., Lukovitch V., Rachkovskij D., Adaptive High Performance Classifier Based on Random Threshold Neurons, Proc. of Twelfth European Meeting on Cybernetics and Systems Research (EMCSR-94), Austria, Vienna, 1994, pp. 1687–1695.
20. Kussul E., Baidyk T., Kasatkina L., Lukovich V., Rosenblatt Perceptrons for Handwritten Digit Recognition, Proceedings of International Joint Conference on Neural Networks IJCNN, 2001, Vol. 2, 2001, pp. 1516–1520.
21. LeCun, Y., Bottou, L., Bengio, Y., Haffner, P., Gradient-based Learning Applied to Document Recognition, Proceedings of the IEEE, Vol. 86, No. 11, November 1998, pp. 2278–2344.
22. Hoque M.S., Fairhurst M.C., A Moving Window Classifier for Off-line Character Recognition. Proceedings of the 7th International Workshop on Frontiers in Handwriting Recognition, 2000, Amsterdam, pp. 595–600.
23. S. Belongie, J. Malik, J. Puzicha, Matching Shapes, Proceedings of the 8-th IEEE Intern. Conf. on Computer Vision ICCV, Vol. 1, 2001, pp. 454–461.
24. S. Belongie, J. Malik, J. Puzicha, Shape matching and object recognition using shape contexts, *IEEE Transactions on Pattern Analysis and Machine Intelligence*, Vol. 24, No. 4, 2002, pp. 509–522.
25. L.Bottou, C. Cortes, J. Denker, H. Drucker, L. Guyon, L. Jackel, Y. LeCun, U. Muller, E. Sackinger, P. Simard, V. Vapnik, Comparison of Classifier Methods: A Case Study in Handwritten Digit Recognition, Proceedings of 12th IAPR Intern. Conf. on Pattern Recognition, 1994, Vol. 2, pp. 77–82.
26. E. Kussul, T. Baidyk, Improved Method of Handwritten Digit Recognition Tested on MNIST Database, Proceedings of the 15th Intern. Conf. on Vision Interface, Calgary, Canada, 2002, pp. 192–197.
27. Baidyk T., Kussul E., Makeyev O., Caballero A., Ruiz L., Carrera G., Velasco G. Flat Image Recognition in the Process of Microdevice Assembly, *Pattern Recognition Letters*. Vol. 25/1, 2004, pp. 107–118.
28. Kussul E., Baidyk T., Wunsch D., Makeyev O., Martín A. Permutation Coding Technique for Image Recognition Systems. *IEEE Transactions on Neural Networks*, Vol. 17/6, November 2006, pp. 1566–1579.
29. Cheng-Lin Liu, K. Nakashima, H. Sako, H. Fujisawa, Handwritten Digit Recognition Using State-of-the-art Techniques, Proceedings of the 8th Intern. Workshop on Frontiers in Handwritten Recognition, Ontario, Canada, August 2002, pp. 320–325.
30. Rosenblatt, F., Principles of Neurodynamics, Spartan books, New York, 1962.
31. Kussul E., Baidyk T., Neural Random Threshold Classifier in OCR Application, Proceedings of the Second All-Ukrainian Intern. Conf. UkrOBRAZ'94, Ukraine, December 1994, pp. 154–157.
32. Kussul E., Kasatkina L., Rachkovskij D., Wunsch D., Application of Random Threshold Neural Networks for Diagnostics of Micro Machine Tool Condition, International Joint Conference on Neural Networks, NJ Alaska, 1998, Vol. 1, pp. 241–244.
33. Baidyk T., Kussul E., Makeyev O. Texture Recognition with Random Subspace Neural Classifier, *WSEAS Transactions on Circuits and Systems*, Issue 4, Vol. 4, April 2005, pp. 319–325.

Chapter 4
Permutation Coding Technique for Image Recognition System

A feature extractor and neural classifier for a face image recognition system are proposed. They are based on the Permutation Coding technique, which continues our investigation of neural networks. The permutation coding technique makes it possible to take into account not only detected features, but also the position of each feature in the image. It permits us to obtain a sufficiently general description of the image to be recognized. Different types of images were used to test the proposed image recognition system. It was tested on the handwritten digit recognition problem and the face recognition problem. The results of this test are very promising. The error rate for the MNIST database is 0.44%, and for the ORL database it is 0.1%. In the last section, which is devoted to micromechanics applications, we will describe the application of the permutation coding technique to the micro-object shape recognition problem.

4.1 Special- and General-Purpose Image Recognition Systems

We have developed a new image recognition method for various applications, based on discovering points of interest on the image and extracting local features of the image around these points. The extracted features are coded in binary form. The binary code of each feature contains information about the localization of this feature in the image, but the code is insensitive to small displacements of the feature in the image. The proposed method recognizes objects having small displacements and small distortions in the image. It could be applied to grayscale or color images. We proved this method on the handwritten digit recognition problem and the face recognition problem.

Handwritten digit recognition could be used for automatic reading of bank checks, customs declarations, postal addresses, and other documents. To test the performance of handwritten digit recognition systems, some publicly available databases are used. Among them, the MNIST database has the following advantages: it contains a large number of samples in training and test sets, and all the

DOI 10.1007/978-3-642-02535-8_4, © Springer-Verlag Berlin Heidelberg 2010

images are centered and transformed to the gray scale. Many recognition systems were tested on this database, and the results are available in the literature [1–7]. Some systems showed good results. The best result in [1] is 70 errors from 10,000 samples using the convolution neural network LeNet-4. In 2001, Belongie [5] reported the result of 63 errors with a system based on the shape matching method [6]. In 2001, we developed the neural classifier LIRA that showed 79 errors [7]. In 2002, we improved the LIRA [8] and repeated Belongie's result of 63 errors (Part II). Further improvement of LIRA classifier permits us to obtain 55 errors [24]. The new recognition system that we proposed gave the result of 44 errors. Cheng-Lin Liu et al. in 2002 investigated handwritten digit recognition by combining the eight methods of feature extraction with seven classifiers. They obtained 56 results for each tested database, including the MNIST database. The best result obtained on the MNIST database was 42 errors [9]. This new method of image recognition based on Permutation Coding Neural Classifier (PCNC) shows the result of 49 errors, 0.49% as a mean value of three runs. The best result of this method is 44 errors.

Automatic face recognition could be used in different security systems (for buildings, banks, etc.), as well as for passport and other document verification. Different approaches are investigated and proposed for solving this task [10–20]. They were tested on the database ORL (Olivetti Research Laboratory) [19, 21–23]. We have tested our classifier on this database and obtained one of the best results. The error rate was 0.1%.

The problem of micro-object shape recognition appears in microequipment technology [46], where we use adaptive algorithms to improve the control systems of micromachine tools. We will describe the results of microscrew shape recognition in the last part of the book. For all the listed problems, we apply the same PCNC algorithm.

At present, very good results are obtained using special recognition systems. Let us, for example, compare the best results obtained in handwritten digit recognition on the MNIST database and face recognition on the ORL database. The MNIST database contains 60,000 handwritten digits in the training set and 10,000 handwritten digits in the test set. There are different classifiers that have been applied in the task of handwriting recognition [1, 2, 5, 6, 8, 9, 24]. The results are presented in Table 3.1.

The ORL database contains 400 photos of 40 persons (ten photos of each person) that differ in illumination, facial expression, and position. Five photos of each person are used for training and the other five photos are used to test the recognition system. The best results obtained on the ORL database are given in Table 4.1 from [25]. As can be seen, almost all of the best recognition systems for the ORL database differ from the recognition systems for the MNIST database. Some of the systems use the same types of classifiers, for example, SVM and multilayer neural networks, but the features extracted from the images in these cases are different.

The great variety of recognition systems takes a huge amount of human work for software development and complicates the development of special hardware that

Table 4.1. Comparative results on the ORL database

Method	Error	Ref.	Year
PDNN	4.0%	[26]	1997
SVM + PCA coef.	3.0%	[27]	2001
Continuous n-tuple classifier	2.7%	[28]	1997
Ergodic HMM + DCT coef.	0.5%	[29]	1998
Classifier with permutation coding	0.1%	[36]	2004
Pseudo 2D HMM + DCT coef.	0%	[30]	1999
Wavelet + HMM	0%	[25]	2003

could ensure high-speed and low-cost image recognition. Therefore, it is necessary to search for more general methods that would give sufficiently good results in different recognition problems. There are general-purpose recognition systems, for example, LeNet [1, 2], Neocognitron [31–34], and Receptive Fields [35], but the recognition quality of such systems is lower than that of special systems. It is necessary to develop a general-purpose recognition system having quality comparable to that of specialized systems.

Some years ago, we started to investigate a general purpose image recognition system. In this system, we use the well-known one-layer perceptron as a classifier. At the center of our investigations is the creation of a general-purpose feature extractor, which is based on the concept of random local descriptors (RLDs). We consider the terms "feature" and "descriptor" as synonyms. We intend to develop the method of RLD creation, which could be used successfully for different types of images (handwriting, faces, vision-based automation, etc.)

4.2 Random Local Descriptors

RLDs are based on two known ideas. The idea of random descriptors of images was proposed by Frank Rosenblatt. In his three-layered perceptron [37], each neuron of the associative layer plays the role of random descriptor of the image. Such a neuron is connected to points randomly selected on the retina (input image) and calculates a function from the brightness of these points. It is important to note that the connections of the neuron are not modifiable during training. Another idea of local descriptors is drawn from the discovery of Hubel and Wiesel [38, 39], who have proved that, in the visual cortex of animals, there are local descriptors that correspond to local contour element orientation, movement, and so on. The discovered local descriptor set is probably incomplete because, in the experiments of Hubel and Wiesel, only those descriptors or features were detected that were initially prepared to present to the animals. Probably, not all the descriptors (features) that can be extracted by the visual cortex of the animals were investigated. Rosenblatt's random descriptors could overcome this drawback, but the application of these descriptors to full-size images decreases the effectiveness of the application.

RLDs are similar to Rosenblatt's descriptors but are applied to the local area of the image. In the first version of our recognition system, each random detector was applied to its own local area, selected randomly in the image. This recognition system was named LIRA. LIRA was tested on the MNIST database and showed sufficiently good results - 55 errors (see Table 3.1). One of the LIRA drawbacks is the sensitivity to image displacement. We compensated this with distortions of input images during the training process; however, this method cannot be used for large displacements. For this reason, we developed a more sophisticated image recognition system.

4.3 General Purpose Image Recognition System Description

The scheme of the general-purpose image recognition system is shown in Fig. 4.1. The base of this system is a multilayer neural network. The first layer, S (sensor layer), corresponds to the input image. The second layer, D_1, contains RLDs of the lowest level, while the layer D_2 contains RLDs of the highest level. The associative layer A contains associative elements, which could be represented by groups of neurons, and the layer R contains the output neurons. Each of these neurons corresponds to the image class under recognition. The scheme of the lowest-level RLD is presented in Fig. 4.2.

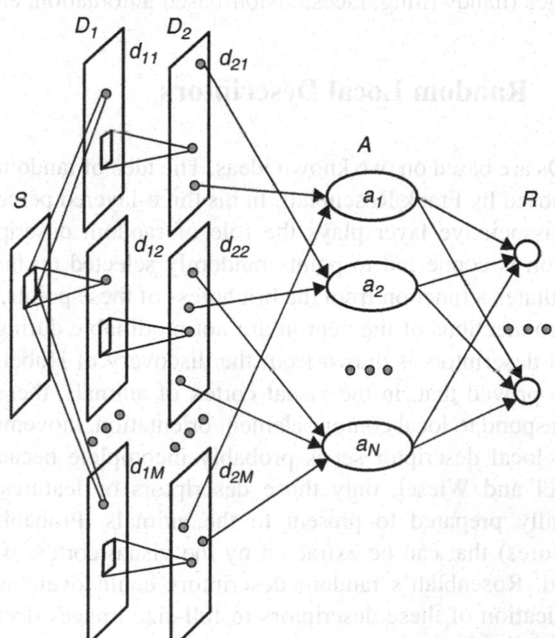

Fig. 4.1 Structure of a general-purpose image recognition system

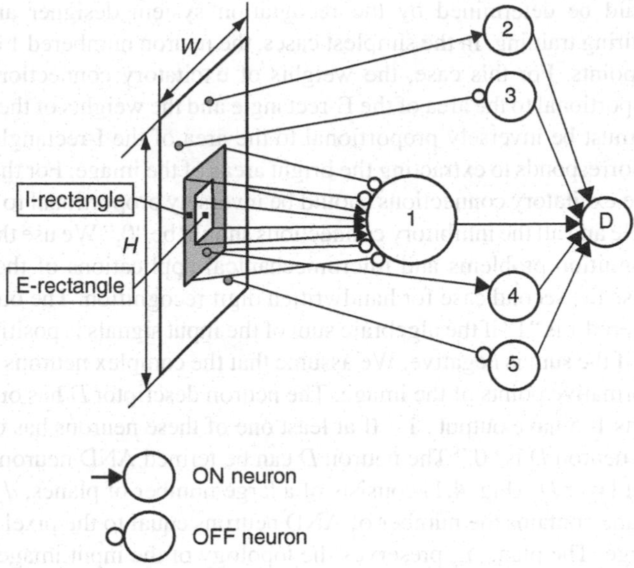

Fig. 4.2 Scheme of the lowest-level RLD

Each RLD contains several neurons (the neurons with numbers 1–5 in Fig. 4.2). All neurons of the RLD are connected to the local area of the *S*-layer (the rectangle with size *H* x *W*). The neurons numbered 2–5 serve for testing the pixels of the *S*-layer, randomly selected in the rectangle. These neurons we call simple neurons, of which there are two types: ON neurons and OFF neurons (similar to ON and OFF neurons of natural neural networks). The outputs of simple neurons are "0" or "1."

The ON neuron has the output "1" if the brightness b_i of the corresponding pixel is higher than the neuron threshold T_i:

$$b_i \geq T_i. \tag{4.1}$$

The OFF neuron has the output "1" if the brightness b_i of the corresponding pixel is less than the neuron threshold T_i:

$$b_i < T_i. \tag{4.2}$$

In Fig. 4.2, the neurons 2 and 4 are ON neurons and the neurons 3 and 5 are OFF neurons. The thresholds T_i are randomly selected in the dynamic range of the input image brightness.

The neuron numbered 1 is the complex neuron. It has excitatory connections with all the pixels from the small rectangle (E-rectangle in Fig. 4.2) located in the center of the rectangle of dimension *H* x *W*, and inhibitory connections with the pixels located around the rectangle (I-rectangle in Fig. 4.2). The weights of the connections could be different in different image recognition problems. These weights should be determined by the recognition system designer and are not modifiable during training. In the simplest cases, the neuron numbered 1 can extract the contour points. For this case, the weights of excitatory connections must be inversely proportional to the area of the E-rectangle and the weights of the inhibitory connections must be inversely proportional to the area of the I-rectangle. Another simple case corresponds to extracting the bright areas of the image. For this case, the weights of the excitatory connections should be inversely proportional to the area of the E-rectangle and all the inhibitory connections should be "0." We use the first case for face recognition problems and micromechanical applications of the proposed system; we use the second case for handwritten digit recognition. The output of the neuron numbered 1 is "1" if the algebraic sum of the input signals is positive, and the output is "0" if the sum is negative. We assume that the complex neurons will detect the most informative points of the image. The neuron descriptor *D* has output "1" if all the neurons 1–5 have output "1." If at least one of these neurons has output "0," the output of neuron *D* is "0." The neuron *D* can be termed AND neuron.

The neural layer D_1 (Fig. 4.1) consists of a large number of planes, $d_{11}, d_{12}, \ldots, d_{1M}$. Each plane contains the number of AND neurons equal to the pixel number of the input image. The plane d_{1j} preserves the topology of the input image, i.e., each neuron of plane d_{1ji} corresponds to the pixel located in the center of the corresponding rectangle of dimension *W* x H (Fig. 4.2). The topology of connections between the sensor layer and neurons 2–5 (Fig. 4.2) is the same in the range of every plane d_{1j} (Fig. 4.1). The topology of connections between the sensor layer and the neuron numbered 1 is the same for all the neurons in all the planes of layer D_1. The aim of each plane is to detect the presence of one concrete feature in any place of the image. The number of planes corresponds to the number of extracted features (in our system, each feature corresponds to one descriptor type). The large number of features permits us to obtain a good description of the image under recognition. In the MNIST database we used 12,800 features; in the ORL database we used 200 features. To estimate the required number of features, it is necessary to solve the problem of structure risk. It is difficult to obtain an analytical solution to this problem for such a complex recognition system as we propose, so we estimate the required number of features experimentally for each recognition problem.

Layer D_2 (Fig. 4.1) also contains *M* planes of neurons. Each neuron of the d_{2j} plane is connected to all the neurons of the d_{1j} plane located within the rectangle. The output of each neuron is "1" if at least one of the connected d_{1j} neurons has the output "1." Such a neuron is termed an OR neuron. The topology of the d_{2j} neurons corresponds to the topology of the d_{1j} neurons and to the topology of the *S* layer. We shall term all neurons having output "1" the active neurons.

The associative layer *A* contains *N* associative elements, which collect the activities of D_2 neurons selected randomly. The other function of the associative elements is to reduce the number of active neurons up to the predetermined value. We refer to this function of the associative layer as normalization. This layer also performs the binding function [40, 41]. The structure of the three associative elements a_1, a_2, and a_3 is presented in Fig. 4.3 in the form of electronic circuits, but it could be implemented as a neural network.

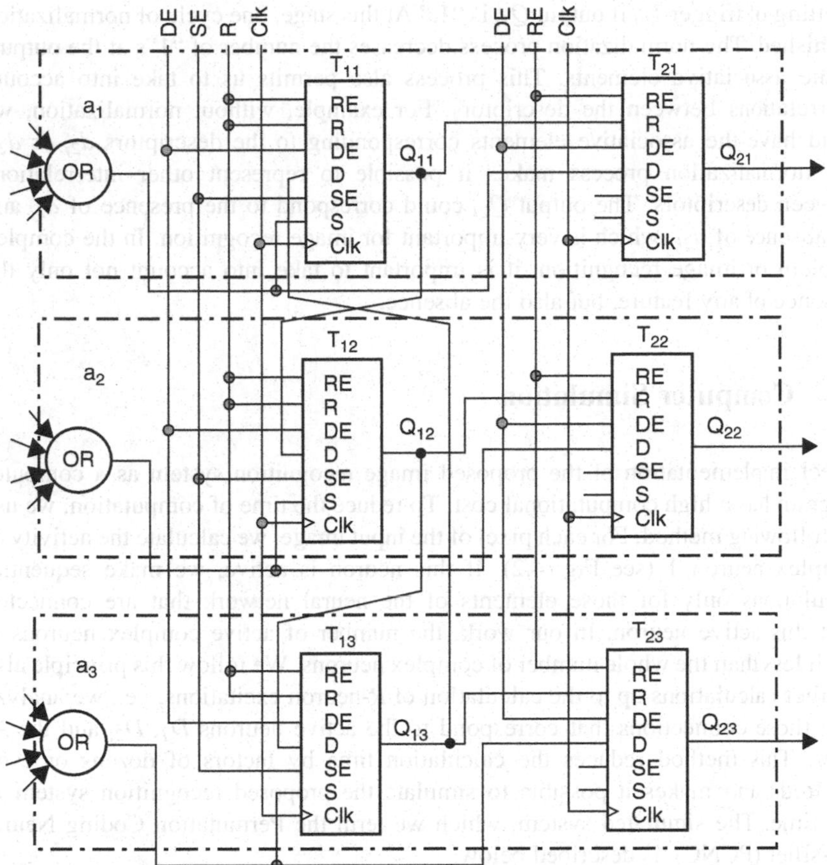

Fig. 4.3 Structure of three associative elements

All the associative elements have the same internal structure, so we describe just the first associative element as an example. The associative element contains at the input an OR neuron connected randomly to several neurons of the D_2 layers. The output of the OR neuron is connected to the S-input of the D-trigger T_{11} and to the D-input of the trigger T_{21}. The output of trigger T_{11} is connected to the D-input randomly selected in the same layer of triggers (with T_{12} in Fig. 4.3). The output of the trigger T_{21} serves as the output of the associative element.

The associative element works as follows. At the beginning, a reset pulse is applied to the trigger T_{11}. After that, a clock pulse is applied to the triggers T_{11} and T_{21}. After these pulses, the outputs Q_{11} and Q_{21} are equal to the output of the OR neuron. After this procedure, the process of normalization begins, which consists of several cycles. Each cycle starts from the pulses applied to the inputs Data Enable (DE) and Clock (Clk) of the trigger T_{11}. These pulses lead to permutations of the outputs of the triggers T_{1i}. Reset Enable (RE) and Clock (Clk) of trigger T_{21} leads to resetting of trigger T_{2i} if output Q_{1i} is "1." At this stage, one cycle of normalization is finished. The normalization process decreases the number of "1"s at the outputs of the associative elements. This process also permits us to take into account interrelations between the descriptors. For example, without normalization, we could have the associative elements corresponding to the descriptors d_{2j} or d_{2k}. The normalization process makes it possible to represent other interrelations between descriptors. The output Q_{2i} could correspond to the presence of d_{2j} and the absence of d_{2k}, which is very important for image recognition. In the complex problem of image recognition, it is important to take into account not only the presence of any feature, but also the absence.

4.4 Computer Simulation

Direct implementation of the proposed image recognition system as a computer program has a high computational cost. To reduce the time of computation, we use the following method. For each pixel of the input image, we calculate the activity of complex neuron 1 (see Fig. 4.2). If this neuron is active, we make sequential calculations only for those elements of the neural network that are connected with this active neuron. In our work, the number of active complex neurons is much less than the whole number of complex neurons. We follow this principle also in other calculations up to the calculation of R-neuron excitations, i.e., we analyze only those connections that correspond to the active neurons D_1, D_2, and the A-layer. This method reduces the calculation time by factors of dozens or even hundreds and makes it possible to simulate the proposed recognition system in real time. The simulated system, which we term the Permutation Coding Neural Classifier (PCNC), is described below.

4.5 Permutation Coding Neural Classifier (PCNC)

We worked out the principles of the proposed method on the basis of the associative-projective neural networks (APNN) [42, 43]. We use the APNN paradigm as a generic structure for different applications, such as random threshold classifier (RTC) [44, 45] random subspace neural classifier, LIRA [7, 8], and PCNC [46, 47].

Here, we propose a new algorithm that outperforms its predecessors due to new elements (feature extractor and encoder) included in its structure.

4.5.1 PCNC structure

The PCNC structure is presented in Fig. 4.4. The image is input to the feature extractor, and the extracted features are applied to the encoder input. The encoder

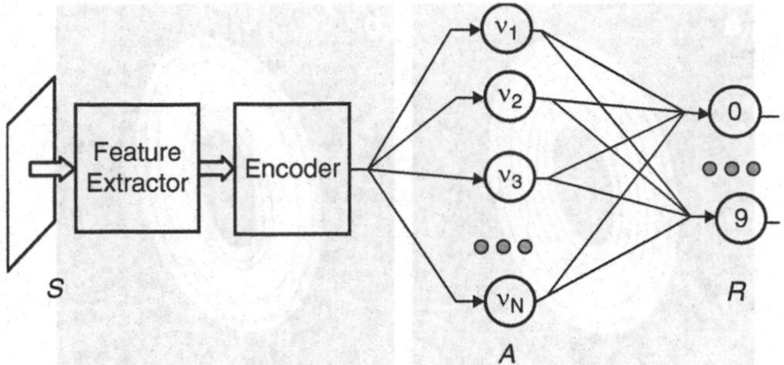

Fig. 4.4 Structure of the Permutation Coding Neural Classifier

produces an output binary vector of large dimension, which is presented to the one-layer neural classifier input. The output of the classifier gives the recognized class.

4.5.2 Feature extractor

In Fig. 4.5*a*, a grayscale image of the digit "0" is shown. The feature extractor begins the work by selecting the points of interest in the image. In principle, various methods of selecting these points could be proposed. For example, the contour points could be selected as the points of interest. For handwritten digit recognition, we selected the points of interest P_{ij}, which have the brightness b_{ij} higher than the predetermined threshold B. These points correspond to the white area in Fig. 4.5b.

The rectangle of area $h*w$ is formed around each point of interest (Fig. 4.6). Multiple features are extracted from the image in the rectangle. The p positive and the n negative points determine each feature. The positive point is the point of the S-layer connected to the ON neuron (Fig. 4.2). The negative point is the point of the S-layer connected to the OFF neuron (Fig. 4.2).

These positive and negative points are randomly distributed in the rectangle. Each point P_{mk} has the threshold T_{mk} that is randomly selected from the range:

$$T_{min} \leq T_{mk} \leq T_{max} \tag{4.3}$$

The feature exists in the rectangle if all its ON and OFF neurons (Fig. 4.2) are active. In other cases, the feature is absent in the rectangle. We used a large number of different features F_i ($i = 1, \ldots, S$). In the final experiments, we worked with $p = 3$; $n = 6$; $h = w = 11$; $T_{min} = 1$; $T_{max} = 254$; and $B = 127$. The feature extractor examines all S features for each point of interest. In the handwritten digit image, the number of the S features was changed from 1,600 to 12,800.

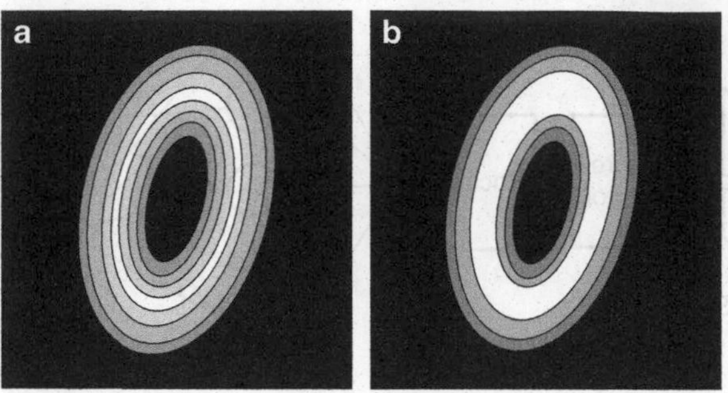

Fig. 4.5 Grayscale image of “0”

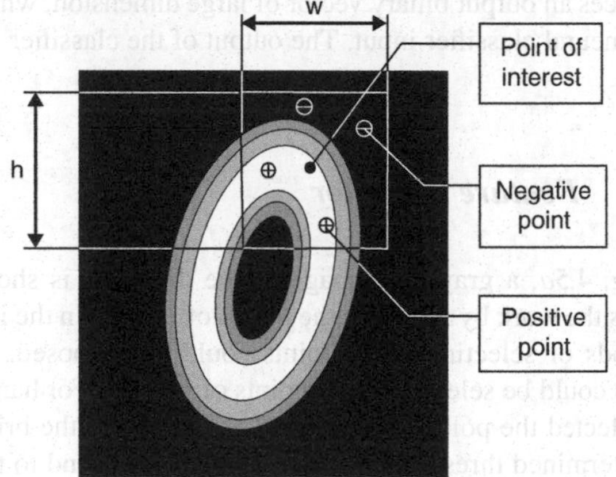

Fig. 4.6 Points of interest selected by the feature extractor (black area)

In the MNIST database, the images have dimensions of 28 x 28 pixels. For the feature extractor, we expanded the images to $h/2$ at the top and bottom sides and to $w/2$ at the left and right sides. All extracted features are transferred to the encoder.

4.5.3 Encoder

The encoder transforms the extracted features to the binary vector:

$$V = \{v_i\}(i = 1, K, N) \tag{4.4}$$

where $v_i = 0$ or 1. For each extracted feature, F_s, the encoder creates an auxiliary binary vector:

$$U = \{u_i\}(i = 1, K, N) \tag{4.5}$$

where $u_i = 0$ or 1. This vector contains K 1s, where $K << N$. In our experiments, K = 16, and the number of neurons N was changed from 32,000 to 512,000. The positions of 1s in the vector U_s are selected randomly for each feature F_s. This procedure generates the list of the positions of 1s for each feature and saves all such lists in the memory. We term the "mask" of the feature F_s the vector U_s.

In the next stage of encoding, it is necessary to transform the auxiliary vector U to the new vector U^*, which corresponds to the feature location in the image. This transformation is made with permutations of the vector U components (Fig. 4.7). The number of permutations depends on the feature location on the image. The permutations in horizontal (X) (Fig. 4.7a) and vertical (Y) (Fig. 4.7b) directions have different permutation schemes and are applied to the vector U sequentially. Firstly, we have to apply X permutations, which give the vector U'', and then apply Y permutations to the vector U'' to obtain the resulting vector U^*.

The problem is to obtain such binary codes of the features that are strongly correlated if the distance between the feature locations is small and those that are weakly correlated if the distance is large. For example, if the feature F_s is extracted at the top point of the handwritten digit and the same feature is extracted at the bottom point of the digit, they must be coded by different binary vectors U^*_{s1} and U^*_{s2}, that have a weak correlation. If the same features are extracted at the neighboring points, they must be coded with almost the same vectors U^*_{s3} and U^*_{s4}. This property makes the recognition system insensitive to the small displacements of the digits in the image.

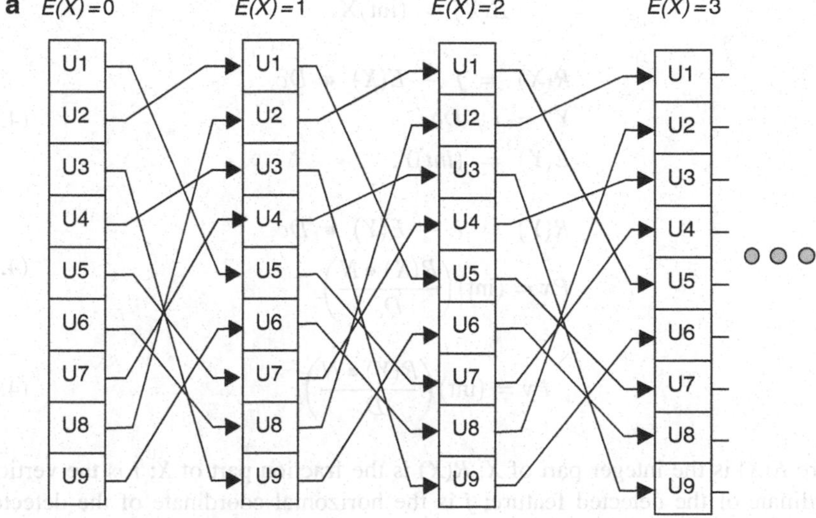

Fig. 4.7a Permutation pattern for X coordinate

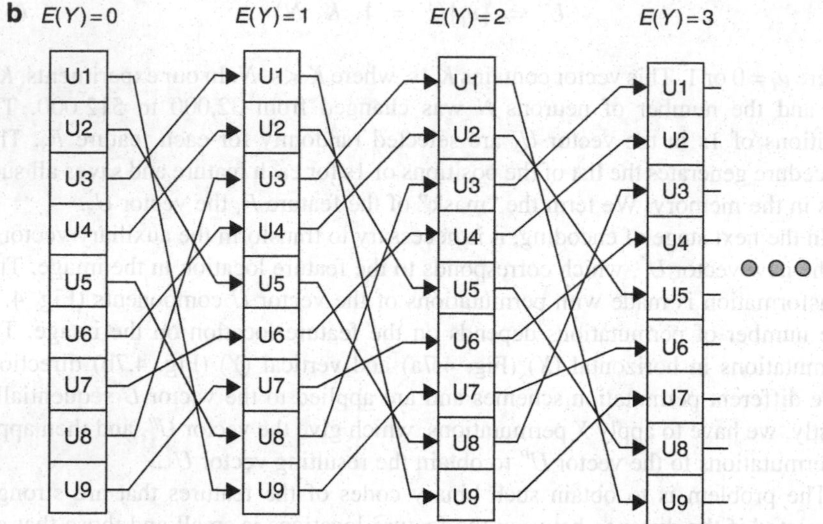

Fig. 4.7b Permutation pattern for *Y* coordinate

To code the feature F_s location in the image, it is necessary to select the correlation distance D_c. Let the same feature F_s be detected in two different points, P_1 and P_2. If the distance d between them is $< D_c$, the corresponding codes will be correlated. If the distance $d > D_c$, the codes will be uncorrelated. To obtain this property, we have to calculate the following values:

$$X = j / Dc, \quad E(X) = (\text{int})X, \tag{4.6}$$

$$R(X) = j - E(X) \bullet Dc, \quad Y = i / Dc, \quad E(Y) = (int)Y, \tag{4.7}$$

$$R(Y) = i - E(Y) \bullet Dc, \quad Px = (\text{int})\left(\frac{R(X) \bullet N}{D_c}\right) \tag{4.8}$$

$$Py = (\text{int})\left(\frac{R(Y) \bullet N}{D_c}\right), \tag{4.9}$$

where $E(X)$ is the integer part of X; $R(X)$ is the fraction part of X; i is the vertical coordinate of the detected feature; j is the horizontal coordinate of the detected feature, N is the number of neurons; and P_x, P_y are the fractions of the neurons for which an additional permutation is needed.

The mask of the feature F_s is considered a code of the feature; it is located at the top-left corner of the image. To shift the feature location in the horizontal direction, it is necessary to make its permutations $E(X)$ times and to make an additional permutation for P_x components of the vector. After that, it is necessary to shift the code vertically, making the permutations $E(Y)$ times and an additional permutation for P_y components.

Let the feature F_k be detected at the point $j = 11$ and $i = 22$; $D_c = 8$; N = 9. In this case, $E(X) = 1$; $E(Y) = 2$; $P_x = 3$; $P_y = 6$. In Fig. 4.8a and b, all components that have to be permutated are shown in gray. To make the permutations of the *kth* component of the vector U, it is necessary to select the *kth* cell in the first column in Fig. 4.8a, and then to follow the arrows until the first white cell appears. This white cell corresponds to the new position of the selected component, which it is necessary to select in the first column of Fig. 4.8b and to follow the arrows until the first white cell appears. This white cell corresponds to the final position of the selected component. For example, the trajectory of component U_3 will be (Fig. 4.8a, b):

$$U_3 \rightarrow U_9 \rightarrow U_7 \rightarrow U_6 \rightarrow U_4. \tag{4.10}$$

The permutations of all components are shown in Fig. 4.9. All the arrows between columns 1 and 2 shown in Fig. 4.8 are generated randomly with the following procedure:

1. For the component U_1 of column 1, randomly select a component of column 2 (U_5 in Fig. 4.8) and connect U_1 to U_5;

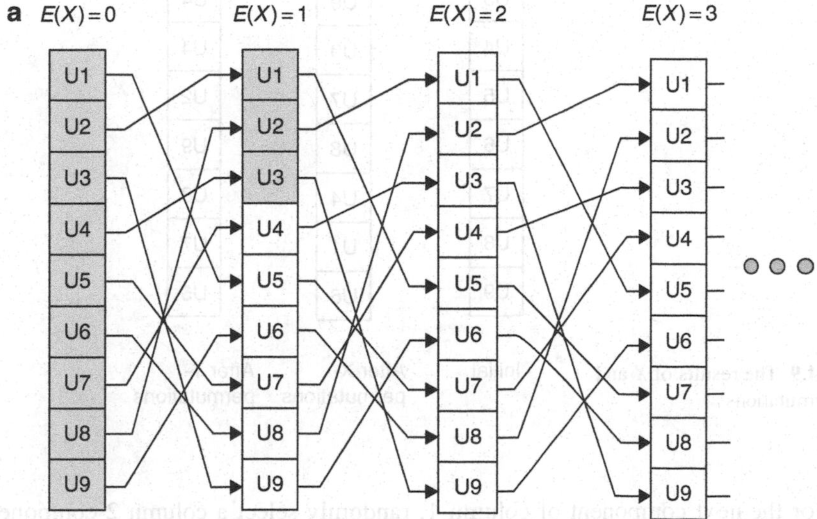

Fig. 4.8a An example of *X* permutations

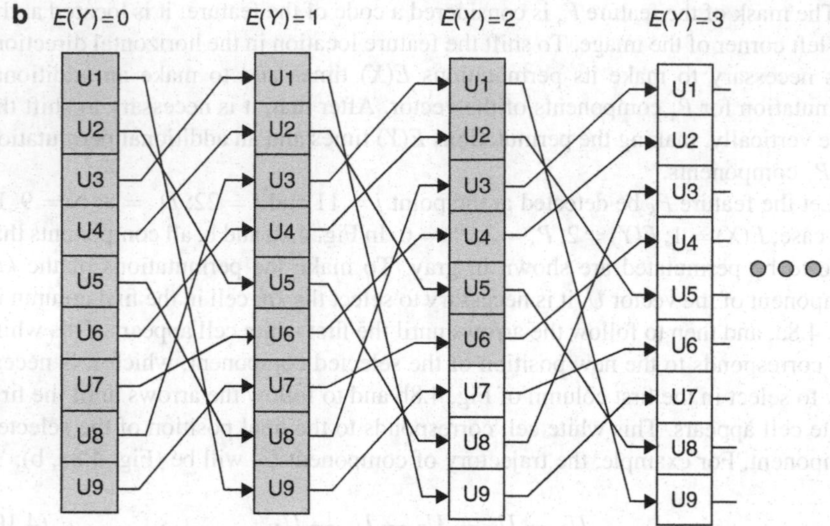

Fig. 4.8b An example of *Y* permutations

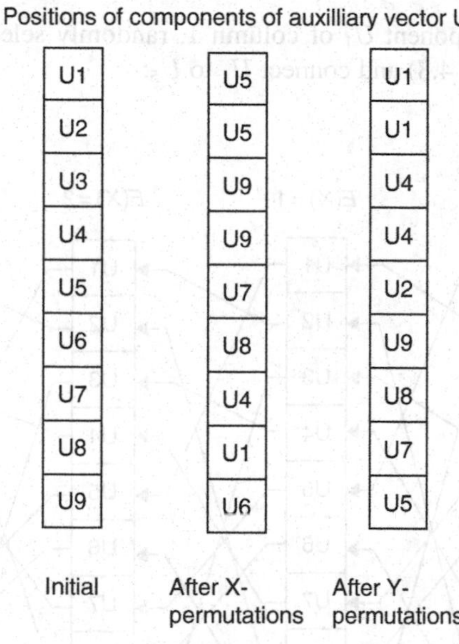

Fig. 4.9 The results of *X* and *Y* permutations

2. For the next component of column 1, randomly select a column 2 component that has no connection and make a new connection between these components;
3. Repeat step 2 until all the components have connections.

The structure of the arrows between the other columns repeats the structure between the first column and the second one.

Let us consider the properties of these permutations. Let the feature F_s be detected in two locations, $P_1(i_1, j_1)$ and $P_2(i_2, j_2)$. Let $dx = |j_2 - j_1|$; $dy = |i_2 - i_1|$. The feature mask corresponds to the vector U. After the X permutations, the vectors U_1 and U_2 will be different. Let Δn be the Hamming distance between U_1 and U_2. The mean value of Δn can be calculated with the approximate equation:

$$\Delta n \approx 2 \cdot K \frac{dx}{D_c}, \text{ if } (dx < D_c),$$

and

$$\Delta n \approx 2 \cdot K, \text{ if } (dx >= D_c). \tag{4.11}$$

After Y permutations, the vectors U_1^* and U_2^* will have Hamming distance Δn_1, which could be estimated as:

$$\Delta n_1 = 2 \cdot K \cdot \overline{\Delta n}, \tag{4.12}$$

where

$$\overline{\Delta n} \approx \left(1 - \left(1 - \frac{dx}{D_c}\right)\left(1 - \frac{dy}{D_c}\right)\right), \text{ if } (dx < D_c) \text{ and } (dy < D_c),$$

and

$$\overline{\Delta n} \approx 1 \text{ if } (dx >= D_c) \text{ or } (dy <= D_c).$$

Thus, vectors U_1^* and U_2^* are correlated only if $dx < D_c$ and $dy < D_c$. The correlation increases if dx and dy are decreasing.

It is seen in Fig. 4.9 that different components of vector U could be placed in the same cell after the accomplished permutations. For example, after the X permutations, the components U_1 and U_2 are allocated in the cells U_5; U_3 and U_4 are allocated in the cell U_9. Such events are undesirable.

Let us consider three cases of component values:

1. $U_1 = 1$; $U_2 = 0$;
2. $U_1 = 0$; $U_2 = 1$;
3. $U_1 = 1$; $U_2 = 1$.

All these cases will give the same result after permutations: $U_5 = 1$. We term this event "absorption of 1s." Absorption leads to the partial loss of information; therefore, it is interesting to estimate the probability of absorption in the permutation process.

Let vector U have N components. K components of the vector have values "1." Let us introduce the two vectors U_a^* and U_b^*. Vector U_a^* contains $R(X) \cdot N$ components, obtained from the initial vector U using permutations $E(X)$ times. All the rest of the components of U_a^* are zero. Vector U_b^* contains $(1 — R(X)) \cdot N$ components, obtained from the initial vector U using permutations $(E(X) + 1)$ times, and the rest of the components are 0.

The probability p_a of "1" in an arbitrary component of vector U_a^* is:

$$p_a = R(X) \cdot \left(\frac{K}{N}\right). \tag{4.13}$$

In the vector U_b, the corresponding probability p_b is:

$$p_b = (1 - R(X)) \cdot \left(\frac{K}{N}\right). \tag{4.14}$$

Absorption occurs if both vectors U_a^* and U_b^* have "1" in the same place. The probability of this event for one component is:

$$p = p_a \cdot p_b = R(X) \cdot (1 - R(X)) \cdot \left(\frac{K^2}{N^2}\right). \tag{4.15}$$

It is easy to show that we have the maximum of p when
$R(X) = (1 - R(X)) = 0.5$.
Thus:

$$P \leq \frac{K^2}{4 \cdot N^2}. \tag{4.16}$$

For all the N components of the vector U, at least one absorption probability shall be:

$$p^* = 1 - (1 - p)^N \leq \left(1 - \frac{K^2}{4 \cdot N^2}\right)^N = e^{-\frac{K^2}{4 \cdot N}}. \tag{4.17}$$

If we have, for example, $K = 16$, $N = 128{,}000$,

$$p^* = 1 - e^{-\frac{1}{2000}} = 0.0005. \tag{4.18}$$

Thus, for a large number of neurons N and a small number of active neurons K, absorption has a very low probability and has practically no influence on the coding process. The code vector V is composed from all code vectors U^*_s of the detected features:

$$v_i = OR_s u^*_{si}, \tag{4.19}$$

where v_i is the i component of the vector V, u^*_{si} is the i component of the vector $U^*{}_s$, which corresponds to the detected feature F_s, and OR_s is disjunction. This coding process produces almost independent representations of all features because we use independent random numbers for the feature mask generation. The weak influence of one feature on the other appears only from the absorption of 1s in disjunction (Equation 4.19).

To recognize the images, it is necessary to use feature combinations. For example, the feature F_a is present in the image, but the feature F_b is absent. To take into account such feature combinations, Context Dependent Thinning (CDT) was proposed [41]. CDT was developed on the basis of vector normalization procedures [48]. There are different procedures of CDT implementation, but here we use the following procedure. The new permutation pattern (Fig. 4.10), which is independent of the X and Y permutations, is generated.

After that, we test each component v_i of the vector V. If $v_i = 0$, nothing must be done. If $v_i = 1$, we consider the trajectory of this component during permutations (according to arrows). If this trajectory contains at least one "1," the value of v_i is converted to 0. For example, in Fig. 4.10, the trajectory of the v_3 component is $v_4 \rightarrow v_8 \rightarrow v_7$. If v_4 or v_8 or v_7 equals 1, we put $v_3 = 0$. The number of the permutations Q in CDT is the parameter of the recognition system. In our experiments, we used Q from 5 to 15. After the realization of CDT, the binary vector V is prepared for the neural classifier.

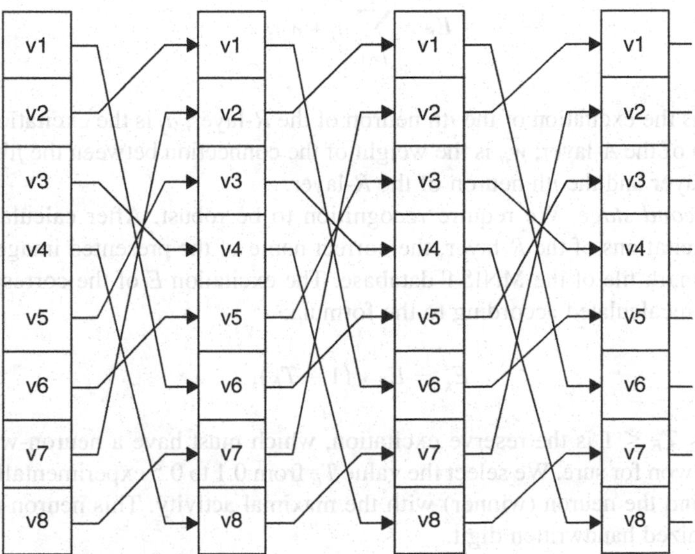

Fig. 4.10 Permutation pattern for Context Dependent Thinning

4.6 PCNC Neural Classifier Training

Earlier, we described the handwritten digit recognition system based on the Rosenblatt perceptron [7, 8]. The one-layer perceptron has very good convergence, but it demands linear separability of the classes in the parametric space. To obtain linear separability, it is necessary to transform the initial parametric space presented by pixel brightness to a parametric space of a larger dimension. In our case, the feature extractor transforms an initial 784-D space into N-dimension space represented by binary code vector V. We examined the systems with N from 32,000 to 512,000. This procedure improves linear separability. The structure of the simplest classifier LIRA, which uses the transformation, was presented in Fig. 3.7. In the classifier LIRA, the associative neuron layer A is connected with the sensor layer S with the randomly selected, non-trainable connections. The set of these connections could be considered as a feature extractor.

In this section, we propose the new recognition system, which contains the feature extractor and the encoder instead of the mentioned connection set (Fig. 4.4). The neural layers A and R are the same as in the previous recognition system. The training rules for connections between the layers A and R also are the same [8]. Let us consider them in more detail. Before training, all the weights of connections between neurons of the A-layer and the R-layer are set to 0.

The first stage. The training procedure begins with the presentation of the first image to the PCNC. The image is coded, and the R-layer neuron excitations E_i are computed. E_i is defined as:

$$E_i = \sum_{j=1}^{N} a_j * w_{ji}, \tag{4.20}$$

where E_i is the excitation of the ith neuron of the R-layer; a_j is the excitation of the jth neuron of the A-layer; w_{ji} is the weight of the connection between the jth neuron of the A-layer and the ith neuron of the R-layer.

The second stage. We require recognition to be robust. After calculating the neuron excitations of the R-layer, the correct name of the presented image is read from the mark file of the MNIST database. The excitation E of the corresponding neuron is recalculated according to the formula:

$$E_k^* = E_k * (1 - T_E), \tag{4.21}$$

where $0 \leq T_E \leq 1$ is the reserve excitation, which must have a neuron-winner to say that it won for sure. We select the value T_E from 0.1 to 0.5 experimentally. After that, we find the neuron (winner) with the maximal activity. This neuron presents the recognized handwritten digit.

The third stage. Denote the neuron-winner number as j, and the number of the neuron, which really corresponds to the input image, as c. If $j = c$, nothing is to be done. If $j \neq c$, the calculations are made according to the equations:

$$\begin{aligned} w_{ic}(t+1) &= w_{ic}(t) + a_i, \\ w_{ij}(t+1) &= w_{ij}(t) - a_i, \text{ if } \left(w_{ij}(t+1) < 0\right),\ w_{ij}(t+1) = 0, \end{aligned} \tag{4.22}$$

where $w_{ij}(t)$ is the weight of the connection between the i-neuron of the A-layer and the j-neuron of the R-layer before reinforcement, $w_{ij}(t+1)$ is the weight after reinforcement, and a_i is the output signal (0 or 1) of the i-neuron of the A-layer.

Taking into account the small number of active neurons, it is convenient to represent the binary vector $\boldsymbol{A}$ not explicitly but as a list of numbers of active neurons. For example, let the vector $\boldsymbol{A}$ be:

$$A = 00010000100000010000. \tag{4.23}$$

The corresponding list of the numbers of active neurons will be 4, 9, and 16. This list is used to save the image codes in compact form and to calculate quickly the activity of the neurons of the output layer. Thus, after executing the coding procedure, every image has a corresponding list of numbers of active neurons.

The training process is carried out iteratively. After representation of all the images from the training subset, the total number of training errors is calculated. If this number is higher than 1% of the total number of images, then the next training cycle is continued. If the error number is less than 1%, the training process is stopped. The process is also stopped when the cycle number is more than a initially prescribed value. In our experiments, we used 30 cycles with the MNIST database and 200 cycles with the ORL database. The flowchart of the algorithm is shown in Fig. 4.11 (Parts 1 and 2).

It is known [3] that the performance of the recognition systems could be improved due to the distortions of the input image during the training process. In [8], we showed that further improvement could be obtained if the distortions are used not only in the training process, but in the recognition process, too.

In the MNIST database, we used only skewing as image distortion. In our experiments, we used three skewings (11.3°, 18.5°, 26.5°) to the left side and three skewings of the same angles to the right side. Thus, the total number of distortions in the training process was six. In the recognition process, we tested three variants of the system: without distortions, with two distortions, and with four distortions. In the ORL database, we used the shifted images for training and did not use distortions for recognition.

4.7 Results Obtained on the MNIST Database

The MNIST database contains 60,000 samples in the training set and 10,000 samples in the test set. The results of the experiments are presented in Table 4.2. The values presented in this table correspond to the number of errors in the recognition of 10,000 test samples. We conducted five experiments with different sizes of the rectangle of dimensions $h * w$ (7 * 7, 9 * 9, 11 * 11, and 13 * 13).

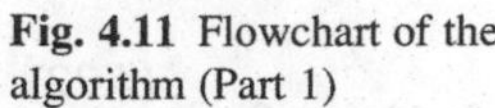

Fig. 4.11 Flowchart of the algorithm (Part 1)

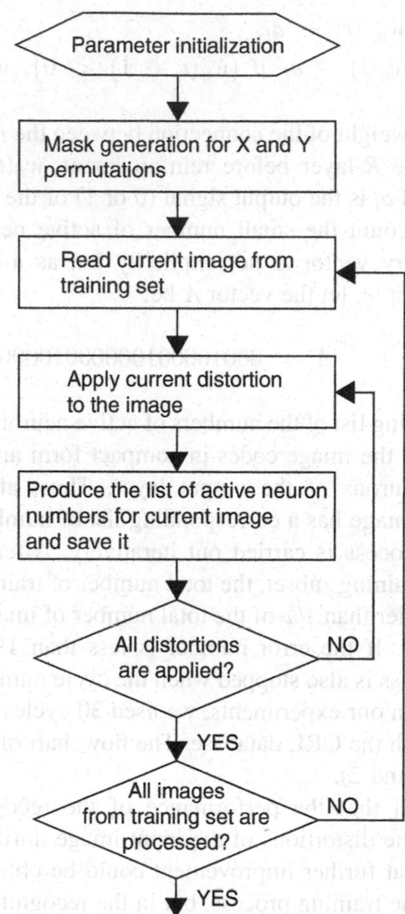

For each size, we made three runs to obtain reliable results, and each run has a different set of random numbers used in equations (4) and (5). For all runs, the parameter T_E was 0.1. The unrecognized digits of the best experiment, in Table 1, line 3, column 11 * 11, are presented in Fig. 4.12.

The experiments utilized a Pentium IV, 3 GHz computer. The training time for 60,000 samples was 55 hours. The recognition time for 10,000 samples was 2,350 seconds (0.235 second per handwritten digit).

4.8 Results Obtained on the ORL Database

The ORL database contains 400 images of 40 persons (ten images of each person). Each image is a full face (Fig. 4.13). The difference between the ten images of each person consists of shifts, head inclinations, facial expressions, and the presence or

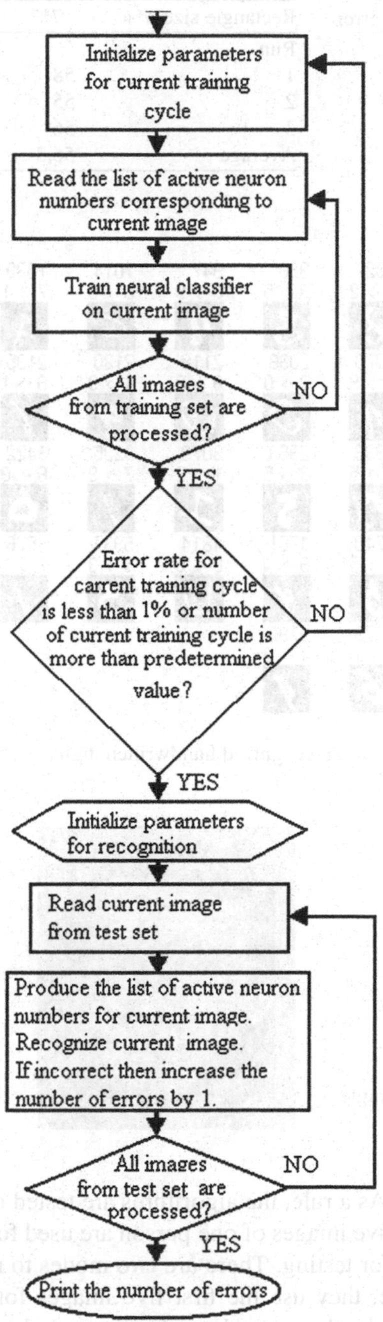

Fig. 4.11 Flowchart of the algorithm (Part 2)

Table 4.2 Recognition error number obtained on the MNIST database

Rectangle size *h***w*	7*7	9*9	11*11	13*13
Run				
1	58	53	56	55
2	55	54	50	51
3	56	54	44	56
Average	56,3	53,7	50	54

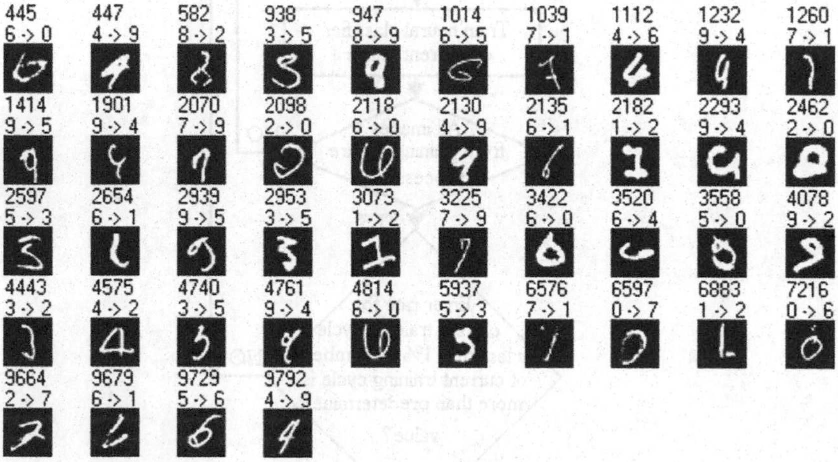

Fig. 4.12 The whole set of unrecognized handwritten digits

Fig. 4.13 The face example from the ORL database

absence of glasses. As a rule, the algorithms are tested on the ORL database in the following manner: five images of one person are used for classifier training and the other five are used for testing. There are two modes to make this partition.

In the first mode, they use the first five images for training and the last five images for testing. In the second mode, they randomly select five images for training, and the rest of the images are used for testing. The first mode was good for comparing the classifiers until classifiers with almost a 100% recognition rate appeared. If the error percentage is less than 2–3%, it is difficult to compare the classifiers using this mode.

The second mode permits us to conduct many different experiments with the same database and to obtain statistically reliable results. In our experiments, we use ten runs for each experiment. This method was used also in other works for estimating the mean value of error rate [11]. They used six runs for each experiment. Their results are given in Table 4.3.

It is also interesting to investigate the cases where less than five images are selected for training and the rest for testing. The data from such experiments are presented in [10]. Unfortunately, the data are presented graphically and no table is given. We tried to restore the data from the graphics. The results are given in Table 4.4:

The results of our experiments are presented in Table 4.5:

The note tr./ex. reflects how many images were used for training (tr.) and how many for examination (ex.). The comparison of results shows that our classifier, in most of the cases, gives the best recognition rate. In this task, we used the classifier that has $N = 64{,}000$; $S = 200$. The parameter T_E was 0.5. We mentioned above that different methods of detecting points of interest could be used. For face recognition, we used the points of the contours. For this purpose, the connections from the E-rectangle (Fig. 4.2) must have positive weights, and the connections from the

Table 4.3 Best performances on 6 simulations

Simulation	1	2	3	4	5	6
NOM	2	6	7	2	1	6

(NOM - number of misclassifications or errors)

Table 4.4 Error rates for different number of each person's image presented for training

Number of training images	The best results from [10] (restored from graphics)	Our results
1	17.6	16.1
2	8.8	7.09
3	4.8	2.15
4	2.8	1.4
5	1.2	0.1

Table 4.5 Recognition results for ORL database

tx/ ex.	NOM										Total NOM	E(%)
	Run											
	1	2	3	4	5	6	7	8	9	10		
1/9	64	66	55	45	52	63	72	44	58	60	580	16.1
2/8	14	29	18	18	25	26	19	13	21	44	227	7.09
3/7	6	13	6	0	9	8	10	3	4	1	60	2.15
4/6	3	3	7	2	5	1	0	7	4	2	34	1.4
5/5	0	0	0	0	0	0	0	0	0	2	2	0.1

I-rectangle must have negative weights. One example of the image with points of interest is presented in Fig. 4.14.

A multipurpose image recognition system was developed, which contains a feature extraction subsystem based on the permutation coding technique and one-layer neural classifier. The main advantage of this system is effective recognition of different types of images. The system was tested on handwritten digit recognition, face recognition, and micro-object shape recognition problems (we will describe this task in micromechanics applications). The large number of features used by this system permits us to obtain a good (complete) description of image properties, and for this reason good recognition results could be achieved. All the features are based on random neural network structures, making them applicable to a wide range of image types. The large number of elements of the associative layer permits us to transform the initial image to the linearly separable space. In this space, the one-layer neural classifier works effectively. So, the advantages of the one-layer classifier (simple training rules and fast convergence) make the system good for different applications.

It was shown in [7, 8] that large-scale classifiers having more than 100,000 neurons in the associative layer demonstrate high performance in image recognition. The main drawback of a perceptron-like classifier of this type (for example, LIRA [49]) is its sensitivity to object displacements in the image. In this work, we proposed a new image recognition system, which is insensitive to small displacements of objects. The best result of this system on the MNIST database is 44 errors (Fig. 4.13). We investigated the influence of the neural classifier size on the classifier performance. The performance is nearly twice as good when the number of neurons in the associative layer of the classifier increases from 32,000 to 512,000. The main drawback of this system is its sensitivity to object rotation and scale change. For this reason, it is necessary to use image distortions during the training and recognition processes. Image distortions increase training and recognition time. In our experiments, the training time was 44 hours on a Pentium IV 1.9 GHz computer, and the recognition time was 35 minutes for 10,000 images. To decrease training and recognition time, it is necessary to develop mask generation methods, which permit a decrease of the sensitivity to rotation and scale change of the object in the image. This work will be done in the future.

Fig. 4.14 The face with points of interest

We examined the proposed recognition system also in the task of face recognition. For this purpose we used the ORL database. To obtain statistically reliable results, we selected the samples for classifier training randomly, and we used the rest of the samples for testing. We conducted ten experiments with different numbers of training samples. For five training samples and five test samples of each face, we obtained 0.1% errors. We also tried other combinations of training and test samples and obtained good results for each combination (see Table 4.5).

The proposed recognition system was also tested for the micro-object shape recognition problem. We will discuss this task and the results obtained in the micromechanics section.

References

1. L.Bottou, C. Cortes, J. Denker, H. Drucker, L. Guyon, L. Jackel, Y. LeCun, U. Muller, E. Sackinger, P. Simard, V. Vapnik, Comparison of Classifier Methods: A Case Study in Handwritten Digit Recognition, Proceedings of 12th IAPR Intern. Conf. on Pattern Reco*gnition*, 1994, Vol. 2, pp. 77–82.
2. http://www.research.att.com/~yann/ocr/mnist/index.html
3. Y. LeCun, L. Bottou, Y. Bengio, P. Haffner, Gradient-based Learning Applied to Document Recognition, *Proceedings of the IEEE*, Vol. 86, No. 11, 1998, pp. 2278–2344.
4. M.S. Hoque, M.C. Fairhurst, A Moving Window Classifier for Off-line Character Recognition, Proceedings of the 7th International Workshop on Frontiers in Handwriting Recognition, 2000, pp. 595–600.
5. S. Belongie, J. Malik, J. Puzicha, Matching Shapes, Proceedings of the 8th IEEE International Conference on Computer Vision ICCV, Vol.1, 2001, pp. 454–461.
6. S. Belongie, J. Malik, J. Puzicha, Shape Matching and Object Recognition Using Shape Contexts, *IEEE Transactions on Pattern Analysis and Machine Intelligence*, Vol. 24, No. 4, 2002, pp. 509–522.
7. E. Kussul, T. Baidyk, L. Kasatkina, V. Lukovich, Rosenblatt Perceptrons for Handwritten Digit Recognition. Proceedings of International Joint Conference on Neural Networks IJCNN'01. Washington, D.C., USA, July 15–19, 2001, pp. 1516–1520.
8. E. Kussul, T. Baidyk, Improved Method of Handwritten Digit Recognition Tested on MNIST Database, Proceedings of the 15th Intern Conf. on Vision Interface, Calgary, Canada, 2002, pp. 192–197.
9. Cheng-Lin Liu, K.Nakashima, H.Sako, H. Fujisawa, Handwritten Digit Recognition Using State-of-the-art Techniques, Proceedings of the 8-th International Workshop on Frontiers in Handwritten Recognition, Ontario, Canada, August 2002, pp. 320–325.
10. Teewoon Tan, Hong Yan, Object Recognition Using Fractal Neighbor Distance: Eventual Convergence and Recognition Rates. Proceedings of the 15th IEEE International Conference on Pattern Recognition, Vol. 2, 3–7 Sept. 2000, pp. 781–784.
11. Meng Joo Er, Shiqian Wu, Juwei Lu, Hock Lye Toh, Face Recognition with Radial Basis Function (RBF) Neural Networks, *IEEE Transactions on Neural Networks*, Vol. 13, No. 3, May 2002, pp. 697–710.
12. S. Singh, M. Singh, M. Markou, Feature Selection for Face Recognition Based on Data Partitioning, Proceedings of the 16th International conference on pattern recognition, Vol. 1, 2002, pp. 680–683.
13. R. Javad Haddadnia, Majid Ahmadi, Karim Faez, An Efficient Method for Recognition of Human Faces Using Higher Orders Pseudo Zernike Moment Invariant, Proceedings of the

Fifth IEEE International Conference on Automatic Face and Gesture Recognition (FGR'02), 2002, p. 6.
14. Wu Xiao-Jun, Wang Shi-Tong, Liu Tong-Ming,Yang Jing-Yu, A New Algorithm of Uncorrelated Discriminant Vectors and its Application. Proceedings of the 4th World Congress on Intelligent Control and Automation, June 2002, Shanghai, P. R. China, pp. 340–344.
15. Victor-Emil Neagoe, Armand-Dragos Ropot, Concurrent Self-organizing Maps for Pattern Classification. Proceedings of the First IEEE International Conference on Cognitive Informatics (ICCI'02), 2002, p. 9.
16. Phiasai T., Arunrungrushi S., Chamnongthai K., Face Recognition System with PCA and Moment Invariant Method, The 2001 IEEE International Symposium on Circuits and Systems, 2001, pp. II-165 - II-168.
17. Gan Jun-ying, Zhang You-wei, Mao Shi-yi, Applications of Adaptive Principal Components Extraction Algorithm in the Feature Extraction of Human Face. Proceedings of the International Symposium on Intelligent Multimedia, Video and Speech Processing, Hong Kong, May 2001, pp. 332–335.
18. Bai-ling Zhang, Yan Guo, Face Recognition by Wavelet Domain Associative Memory. Proceedings of the International Symposium on Intelligent Multimedia, Video and Speech Processing, Hong Kong, May 2001, pp. 481–485.
19. Steve Lawrence, C. Lee Giles, Ah Chung Tsoi, Andrew D. Back, Face Recognition: A Convolutional Neural-network Approach. *IEEE Transactions on Neural Networks*, Vol. 8, No. 1, January 1997, pp. 98–113.
20. Guodong Guo, Stan Z. Li, Kapluk Chan, Face Recognition by Support Vector Machines. In: Proceedings of the Fourth IEEE International Conference on Automatic Face and Gesture Recognition, 2000, 28-30 March 2000, pp. 196–201.
21. S.-H. Lin, S.-Y. Kung, and L.-J. Lin, Face Recognition/-Detection by Probabilistic Decision-based Neural Network, *IEEE Trans. Neural Networks*, Vol.8, Jan. 1997, pp. 114–132.
22. V. Brennan and J. Principe, Face Classification Using Multiresolution Principal Component Analysis, in Proceedings of the IEEE Workshop Neural Networks Signal Processing, 1998, pp. 506–515.
23. S. Eickeler, S. Mueller, and G. Rigoll, High Quality Face Recognition in JPEG Compressed Images, in Proceedings of the IEEE Int. Conf. Image Processing, 1999, pp. 672–676.
24. Baidyk T., Kussul E., Makeyev O., Caballero A., Ruiz L., Carrera G., Velasco G. Flat image recognition in the process of microdevice assembly. *Pattern Recognition Letters*. Vol. 25/1, 2004, pp.107–118.
25. Bicego M., Castellani U., Murino V. Using Hidden Markov Models and Wavelets for Face Recognition. Proceedings of the 12th International Conference on Image Analysis and Processing (ICIAP'03), Italy, 2003, pp. 5.
26. Lin S., Kung S., Lin L. Face Recognition/Detection by Probabilistic Decision-Based Neural Network. *IEEE Transactions on Neural Networks*, 8(1), January 1997, pp. 114–131.
27. Guo G., Li S.Z., Kapluk C. Face Recognition by Support Vector Machines. *Image and Vision Computing*, 19 (9-10), 2001, pp. 631–638.
28. Lucas S. Face Recognition with the Continuous n-tuple Classifier. Proceedings of British Machine Vision Conference, September 1997.
29. Kohir V., Desai U. Face Recognition Using DCT-HMM Approach. Workshop on Advances in Facial Image Analysis and Recognition Technology (AFIART), Freiburg, Germany, June 1998.
30. Eickeler S., Miller S., Rigolli G. Recognition of JPEG Compressed Face Images Based on Statistical Methods. *Image and Vision Computing*, 18, March 2000, pp. 279–287.
31. Fukushima K., Neocognitron of a New Version: Handwritten Digit Recognition. Proceedings of the International Joint Conference on Neural Networks, Vol. 2, 2001, pp.1498–1503.
32. Fukushima K. Neural Network Models for Vision. Proceedings of the International Joint Conference on Neural Networks, 2003, pp.2625–2630.

33. Watanabe A., Andoh M., Chujo N., Harata Y. Neocognitron Capable of Position Detection and Vehicle Recognition. Proceedings of the International Joint Conference on Neural Networks, Vol. 5, 1999, pp.3170–3173.
34. San Kan Lee, Pau-choo Chung, et al. A Shape Cognitron Neural Network for Breast Cancer Detection. Proceedings of the International Joint Conference on Neural Networks, Vol. 1, 2002, pp. 822–827.
35. Perez C., Salinas C., Estévez P., Valenzuela P. Genetic Design of Biologically Inspired Receptive Fields for Neural Pattern Recognition. *IEEE Transactions on System, Man, and Cybernetics, Part B: Cybernetics*, Vol. 33, No. 2, April 2003, pp. 258–270.
36. Kussul E., Baidyk T., Kussul M., 2004, Neural Network System for Face Recognition. Proceedings of the IEEE International Symposium on Circuits and Systems, ISCAS 2004, May 23-26, Vancouver, Canada, Vol. V, pp.V-768–V-771.
37. Rosenblatt F. Principles of Neurodynamics. Spartan Books, New York, 1962.
38. Hubel D., Wiesel T. Receptive Fields, Binocular Interaction and Functional Architecture in the Cat's Visual Cortex. *Journal on Physiology* (Lond.), 106 (1), 1962, pp. 106–154.
39. Hubel D., Wiesel T. Receptive Fields and Functional Architecture in Two Nonstriate Visual Areas (18 and 19) of the Cat. *Journal on Neurophy*siology, 28 (2), 1965, pp.229–289.
40. Plate T. Holographic Reduced Representations. IEEE Transactions on Neural Networks, 6, 1995, pp. 623–641.
41. D. Rachkovskij, E. Kussul, Binding and Normalization of Binary Sparse Distributed Representations by Context-depending Thinning. Neural Computation 13, 2001, pp. 411-452.
42. E. M. Kussul, D. A. Rachkovskij, T. N. Baidyk, On Image Texture Recognition by Associative-Projective Neurocomputer. Proc. of the ANNIE'91 conference. "Intelligent engineering systems through artificial neural networks", ed. by C.H. Dagli, S.Kumara and Y.C. Shin, ASME Press, 1991, pp. 453–458.
43. E.M. Kussul, D.A. Rachkovskij, T.N. Baidyk, Associative-Projective Neural Networks: Architecture, Implementation, Applications. Proc. of Fourth Intern. Conf. "Neural Networks & their Applications", Nimes, France, Nov. 4–8, 1991, (EC2 Publishing), pp. 463–476.
44. E. Kussul, T. Baidyk, Neural Random Threshold Classifier in OCR Application", Proc. of the Second All-Ukrainian Intern. Conf., Kiev, Ukraine, 1994, pp. 154-157.
45. E. Kussul, T. Baidyk, V. Lukovitch, D. Rachkovskij, Adaptive High Performance Classifier Based on Random Threshold Neurons. In: R. Trappl (Ed.) Cybernetics and Systems '94 (Singapore: World Scientific Publishing Co. Pte. Ltd., 1994), pp. 1687–1695.
46. Baidyk T., Kussul E., 2004, Neural Network Based Vision System for Micro Workpieces Manufacturing. WSEAS Transactions on Systems, Issue 2, Vol. 3, April 2004, pp. 483–488.
47. Kussul E., Baidyk T., Wunsch D., Makeyev O., Martín A. Permutation Coding Technique for Image Recognition Systems. IEEE Transactions on Neural Networks, Vol. 17/6, November 2006, pp. 1566–1579.
48. S. Artikutsa, T. Baidyk, E. Kussul, D. Rachkovskij, Texture Recognition with the Neurocomputer. Preprint 91-8 of Institute of Cybernetics, Ukraine, 1991, pp. 20 (in Russian).
49. E. Kussul, T. Baidyk, Improved Method of Handwritten Digit Recognition Tested on MNIST Database, *Image and Vision Computing*, Vol. 22/12, 2004, pp. 971–981.

Chapter 5
Associative-Projective Neural Networks (APNNs)

Associative-projective neural networks (APNNs) were proposed by E. M. Kussul [1] and were developed with the participation of T. N. Baidyk and D. A. Rachkovskij [2–10]. These networks relate to those with distributed coding methods [11]. Later we shall discuss them in detail. APNNs make it possible to construct hierarchical systems for information processing.

5.1 General Description of the Architecture

Before examining the structure and the properties of the associative-projective neural networks, let us describe the model of the neuron and the model of the associative neural field, which is the base block of this neural network architecture.

5.1.1 Neuron, the Training Algorithms

The neuron model, which for the sake of brevity we will call the neuron, is the basic element of the neural network. The real biological neuron is a very complex element in which the processes connected with information processing occur, as well as other phenomena, such as the processes that ensure the vital activity of the neuron. The neuron consists of the body of the neuron (soma), dendrites (inputs), and the axon (output). The excitation generated by the neuron is transferred to other neurons by a change in the electric potential. Complex chemical, physical, and physiological processes closely interact with each other. Therefore, the essential simplifications depending on the simulation targets are made during the simulation of biological neurons. If the task of the investigation is the study of electrical and chemical transformations and signal generation, then the neuron is usually described by the set of differential equations that reflect the state of the separate neuron elements (the membrane, soma, dendrites, etc.). Sometimes in this case, the geometric characteristics of the neuron can be taken into account. This model can

DOI 10.1007/978-3-642-02535-8_5,

describe sufficiently well a large part of the information characteristics of the neuron, but it is too complex for the simulation of information processes in the networks, which consist of a large quantity (many thousands) of neurons. Since our target is the study of the properties of neuron assemblies, i.e., the large populations of neurons, it is necessary for us to make the maximum permissible number of simplifications, ensuring only that these simplifications would not lead to the loss of the most interesting properties of neuron assemblies. The neuron model given below, in our opinion, is sufficient to achieve this goal, and at the same time its simplicity makes it possible to facilitate the fulfillment of computerized numerical experiments and the development of the corresponding specialized hardware (neurocomputer).

Let us examine the neuron model, which has one output and many inputs (Fig. 5.1). The signal at the output can have two values, 0 or 1.

If the output signal is equal to 1, then the neuron is excited. We shall distinguish between several types of input: set (*S*), inhibitory (*R*), associative (*A*), training (*Tr*), synchronizing (*C*), and threshold control (*Th*). Let us designate inputs and the output of the neuron with uppercase letters, and signals on them with the corresponding lowercase letters. The set (*S*), inhibitory (*R*), associative (*A*), and synchronizing inputs (*C*) are binary-valued. We shall say that the signal is present at the input if it is equal to "1," and it is absent at the input if it is equal to "0." The inputs *Tr* and *Th* obtain gradual signals whose ranges are $-1 < tr < 1$ and $0 < th < n$, where n is a quantity of associative inputs.

The neuron works as follows. The synchronizing input obtains the signal, which has alternately low (0) and high (1) levels. The signal at the neuron output (q) can change only at the moment when the synchronizing signal passes from the low level to the high. If at this moment, on at least one of the inhibitory input signals, r equals 1, then the signal at the output takes the 0 value ($q = 0$), regardless of what signals enter other inputs. So, inhibitory inputs dominate over all other inputs.

If at the inputs *R* the signals are absent, and on at least one of the inputs *S* the signal is present, then at the neuron output appears ($q = 1$). If the signals at the inputs *S* and *R* are absent, then the value of the output signal is determined in accordance with the expression:

$$q = \begin{cases} 1, & if \sum_{i=1}^{n} a_i w_i > th, \\ 0, & if \sum_{i=1}^{n} a_i w_i \le th, \end{cases} \tag{5.1}$$

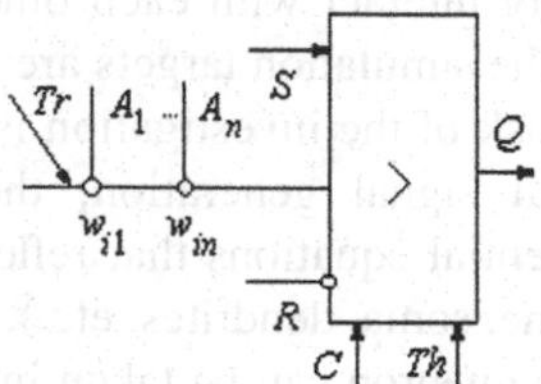

Fig. 5.1. Neuron model

where a_i is the signal at the associative input i of the neuron; w_i is the synaptic weight of the input i of the neuron; and th is the threshold value at the input Th. w_i can take only two values: 0 or 1.

Earlier investigations of neuron assemblies were made by Hebb and Milner [12, 13]. Many of the mechanisms that they proposed were used later by other authors. For example, the biological neural network does not contain primordial knowledge about the world; it obtains knowlege gradually. During training, the neural networks are modified and are capable of memorizing new information. How does training occur, and what processes occur in the neural network? These questions became the subject of consideration in Hebb's book [12]. The fundamental rule (law of training) is formulated by Hebb as follows: if neuron A excites neuron B and this process is repeated, then the synaptic weight of the connection between these neurons increases. This rule of the training of neural networks proposed by Hebb is used frequently and with only small modifications in the newest models of neural networks.

In our case, the modified Hebbian rule of training is used. The training of the neuron is accomplished due to a change of its synaptic weight, which occurs in cases when the signal at the training input is not equal to 0 ($tr \neq 0$). We will distinguish training with the positive signal tr (positive reinforcement) and negative signal tr (negative reinforcement). With positive reinforcement, a change of the synaptic weight is accomplished in accordance with the expression:

$$w_i^* = w_i \mathrm{U}(a_i \;\&\; q \;\&\; h_i), \tag{5.2}$$

where U is a disjunction; & is a conjunction; w_i^* is the synaptic weight after training; w_i is the synaptic weight before training; a_i is the signal at the associative input i; q is the signal at the neuron output; and h_i is the binary random variable, whose probability of unit value is equal to the absolute value of reinforcement tr:

$$p(h_i = 1) = |\,\mathrm{tr}\,|. \tag{5.3}$$

With negative reinforcement, a change in the synaptic weight is described by other expressions:

$$w_i^* = w_i \;\&\; \overline{\left(a_i \;\&\; q\& \; h_i\right)}, \tag{5.4}$$

$$p(h_i = 1) = |tr|, \tag{5.5}$$

where the line above the expression means negation. Such neurons will be used for constructing the associative-projective neuron structures.

The neurons described above will be used in the associative fields. In order to understand the work of the associative field, let us give more detailed ideas about the work of the neuron. Fig. 5.2 depicts the detailed functional diagram of the neuron in which the foregoing designations are used, and new ones are added: *SUM* is the adder, which is followed by the threshold element; *TG* is the flip-flop; Q^* is

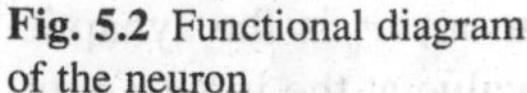
Fig. 5.2 Functional diagram of the neuron

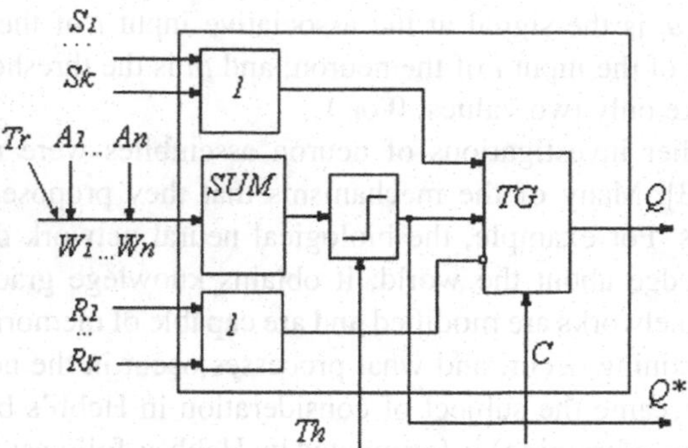

the additional output of the neuron. The set inputs $S_1, \ldots, S_k$ are joined to the OR element. The inhibitory inputs $R_1, \ldots, R_k$ are joined to another OR element. The additional output of neuron Q^* is used to control the activation of the associative field described below.

5.1.2 Neural Fields

The basic structural unit of the proposed associative-projective neural network is a neural field [2, 5]. The neural field is the subset of the neurons, which carry out identical functions. For the neurons of one neural field, all the connections with the neurons of this field and the neurons of other fields are formed according to the rules common for all neurons of the field. We will consider that a quantity of neurons in each neural field equals the same value n. We distinguish between two types of neural fields: associative and buffer. The associative neural field is designated for the formation of neural ensembles during the neural network training and for the associative restoration of neural ensembles in the mode of recognition. The buffer neural field is designated for the temporary storage of ensembles, for the normalization of the sizes of neural ensembles, and for the auxilliary operations with the ensembles. With the aid of the buffer neural field, it is possible to determine the difference between two excited neural ensembles. We also will distinguish between different functions that can realize the buffer field. Any neural field can have the special device, which we will call the counter, that determines a quantity of excited neurons in this field.

Associative neural field.

Let us examine in detail the functions and properties of the associative neural field. The associative field is represented in Fig. 5.3. Like any other field, it contains n neurons, whose outputs are the outputs of the neural field. Furthermore, each of the outputs is supplied to all associative inputs of the neurons of the same field. Thus, the neurons are connected to each other with associative connections. The reinforcement signal tr is supplied to the training inputs of all the neurons of the

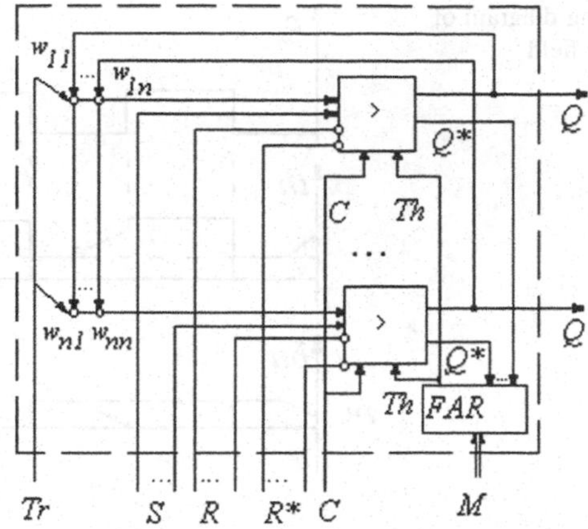

Fig. 5.3 The associative field structure

field. One of the set inputs of each neuron leaves the limits of the field and serves for information input into the associative field (inputs *S*).

We emphasize that the neurons of the associative field are connected with each other with the aid of associative connections, whose weights can change in the process of training. This is important because two additional types of connections will be introduced below: projective and receptive.

Similarly to the set inputs, one of the inhibitory (reset) inputs of each neuron leaves the field and serves for the forced inhibition of excited neurons (inputs *R*). The field has one common input to inhibit all the neurons of the field and the synchronizing input *C*. All the additional outputs of neurons (Q^*) inside the field are supplied to the inputs of the field activity regulator (*FAR*), which determines a quantity of excited neurons (*bn*) and establishes the threshold *th* (common for all neurons) in such a way that the number of active neurons would not exceed a certain value *m*, which is input into the associative field on the input M.

Let us consider the work of the associative field in detail by first examining its timing diagram (Fig. 5.4). The threshold for the associative field neurons must be selected in such a way that no more than *m* neurons in the network will be active. It is achieved in one cycle of the work of the associative field. The duration of the syncronizing signal is selected such that during its first part (high level of the sync signal), the process of summing up the input signals will be completed. The field activity regulator (FAR) begins to work along the back edge of the sync signal. In this case, the outputs *Q* of all neurons of the field are locked and do not change because they can change only along the front edge of the sync signal. The FAR smoothly decreases the threshold *th* (Fig. 5.4) for all neurons of the field and correspondingly changes signals at the additional outputs Q^* so that a quantity of active neurons (marked *bn* in Fig. 5.4) increases. As soon as this quantity of active neurons, *bn*, achieves the defined value *m*, the regulator ceases to change the threshold. As soon as the front edge of the sync signal appears, the pattern of

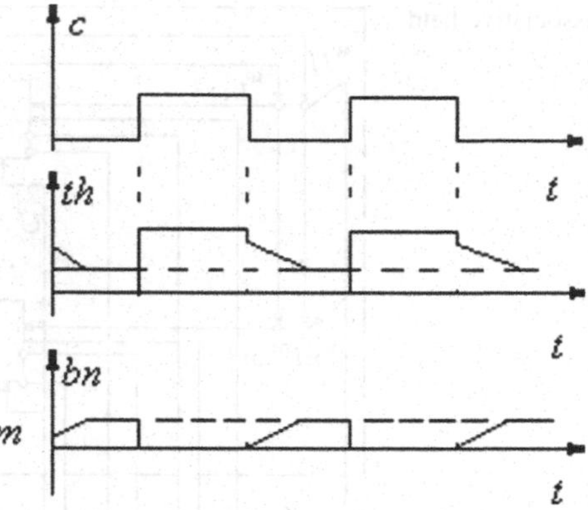

Fig. 5.4 Timing diagram of the associative field

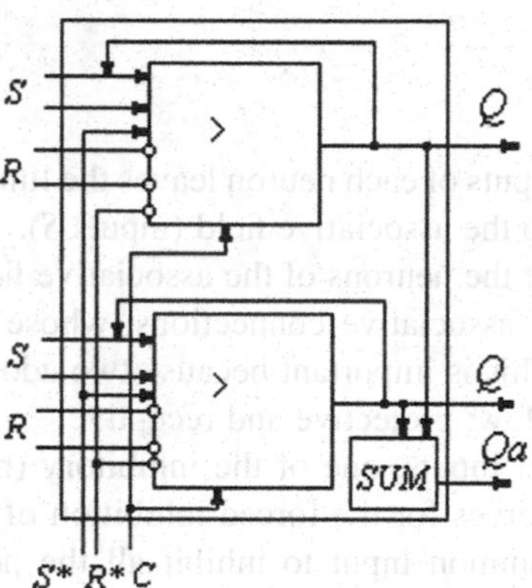

Fig. 5.5 The structure of the buffer field

neuron activity is transferred from the additional outputs Q^* to the basic outputs Q and is fixed for the period of the following sync pulse.

Thus, the regulation of the activity of the associative field is accomplished. A quantity of active neurons of the field always occurs approximately equal to the established value m and, as a rule, only insignificantly exceeds it.

Buffer neural field.

Let us describe the structure of the buffer field (see Fig. 5.5). The neurons of the buffer field do not use the associative inputs, the threshold set input, or the training input. One of the set inputs and one of the inhibitory inputs are common for all neurons of the buffer field (forced installation of neurons into the excited or inhibited state). Synchronous input is common for all the neurons of the buffer field. Additional output in the buffer field is not used. The block *SUM* counts the number of active neurons in the field. The output of each neuron of the buffer field is connected with one of the set inputs of the same neuron, making it possible to accumulate the activity of neurons up to the appearance of a signal of common reset (which inhibits the activity of the field).

Projective and receptive connections.

The projective connections connect the outputs of one neural field with the inputs of another (or the same) neural field, and they do not change during training. The structure of connections can establish the one-to-one correspondence between the neurons of one associative field and those of another associative field that belong to different hierarchical levels. We shall say that two fields are connected with a projective connection if each neuron of one field is connected with a corresponding neuron of another field with a projective connection. The projective connection between the fields can be displaced or undisplaced. In the case of the undisplaced projective connection, the output of the neuron i of the first field is connected with the inputs of the neuron i of the second field. In the case of the displaced projective connection, the output of the neuron i of the first field is connected with the input of $|i + s|_n$ of the second field, where s is the value of the displacement, and $|i + s|_n$ is the sum on the module n.

Projective connections can be excitatory or inhibitory. In the functional diagrams, the projective exciting connection is depicted as a line with an arrow, while the projective inhibitory connection is depicted as a line with a circle. Signals between the fields are transferred by projective connections only in the presence of the resolving signal. If there is no such signal, then the transfer of the signals through the projective connections does not occur. The resolving signal is common for all connections between the neurons of two fields. Projective connections are used for organizing neural fields into complex hierarchical structures to solve different problems of artificial intelligence.

Besides associative connections and projective connections, we will introduce receptive connections. It is necessary to note that in the associative-projective structures oriented toward solving image recognition problems, in which the image is taken from the retina, projective as well as receptive connections are used. Distinctly from projective connections, which map the neuron of one field onto the neuron of another field, receptive connections map the subset of the neurons of one field, called the receptive field (by analogy with the physiology of sight), onto one neuron of another field. Two examples of receptive connections R between the field of retina S_1 and the buffer field B_2 are given in Fig. 5.6.

The receptive fields can have different sizes and shapes depending on the task solved by the system. In order to take this variety into account, it is necessary to provide the methods of their definition. The methods of the definition of receptive fields will determine the methods of the definition of receptive connections.

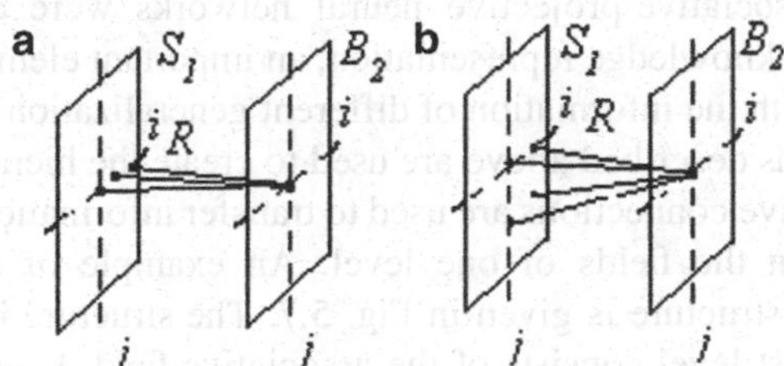

Fig. 5.6 Receptive connections R between different fields

One such method is the direct enumeration of the neurons that belong to this field. Another method of definition is the introduction of functional dependence, which describes the concrete subsets. Then the receptive connections from these fields can be described:

$$R(x_{|i+r_u|_m, |j+s_u|_n} \to y_{ij}(\theta)), \tag{5.6}$$

where x_{ij} are the neurons belonging to the receptive field; y_{ij} are the neurons switched on by the receptive connections from the receptive fields; r_u, s_u are the parameters that determine the position of neurons in the receptive field of the retina; θ is the threshold of the neuron; and m, n are the sizes of the retina. After determining the set of pairs (r_u, s_u), we determine the geometry of the receptive field, so the formula would be valid for all points. The toroidal closing is done on the boundaries of the neural fields (retina).

Let us consider several examples of the receptive connection definition. Figure 5.6 gives an example of receptive fields. In Fig. 5.6a, the receptive field is defined as:

$r_0 = 0, s_0 = 0,$
$r_1 = 1, s_1 = 0,$
$r_2 = 2, s_2 = 0.$

In Fig. 5.6b, the receptive field is described as:

$r_0 = 0, s_0 = 0,$
$r_1 = 0, s_1 = -1,$
$r_2 = 1, s_2 = 0,$
$r_3 = 1, s_3 = -1.$

The receptive connections R reflect the fields defined above into the neurons that belong to the buffer field B_2. The receptive fields can be defined with the aid of the formulas. Thus, for example, if the receptive field is the circle of the specific radius, then it is possible to define it as follows: all neurons for which ${r_u}^2 + {s_u}^2 < r^2$ belong to the receptive field, where r is the defined circle radius. If the receptive field is a ring that has the average radius r and width 2 * d, then its description is:

$$(r-d)^2 < r_u^2 + s_u^2 < (r+d)^2. \tag{5.7}$$

The hierarchical organization of associative-projective neural networks.

The associative-projective neural networks were constructed to be used in systems of knowledge representation, an important element of which is their ability to work with the information of different generalization levels. The associative and buffer fields described above are used to create the hierarchicial system. Projective and receptive connections are used to transfer information from one level to another or between the fields of one level. An example of a hierarchical associative-projective structure is given in Fig. 5.7. The structure includes two levels, L_1 and L_2. The first level consists of the associative field A_1 and the buffer fields B_1, B_2.

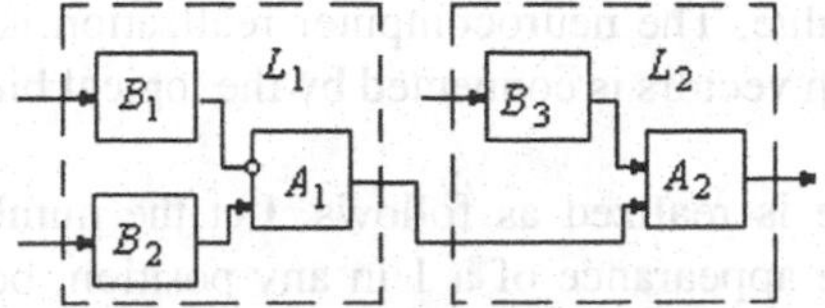

Fig. 5.7 Example of hierarchical associative-projective structure

The associative field A_2 and the buffer field B_3 relate to the level L_2. Inside the associative fields exist the associative connections (they are not shown in Fig. 5.7), but between the fields, the excitatory and inhibitory projective connections are shown.

5.2 Input Coding and the Formation of the Input Ensembles

To describe the activity of the neural field, we introduce the binary vector $V(v_1, \ldots, v_n)$. If in field A the neuron a_i is active, we put $v_i = 1$, and if a_i is not active, we put $v_i = 0$. We say that the vector V is the code of the activity of neural field A.

Many different methods of input information coding for neural networks have been developed and proposed [14–16], which are connected with the great variety of neural networks. In this section, we will consider the so-called stochastic or local connected coding [2, 17, 18], and also shift coding [8]. These coding methods relate to distributed coding with a low level of activity (sparse coding) [11, 19]. This means that in the neural network, any coded object is represented not with one neuron, but with many neurons. Let us term these neurons neuron ensembles. The low level of activity means that the number of neurons in the neuron ensemble is much less than the total number of neurons in the neural field.

5.2.1 Local Connected Coding

The special feature of local connected coding is that the different numerical values of the parameters and the names of these parameters are represented by different binary stochastic vectors. At the input of the associative field, each feature must be represented in the form of the n-dimensional binary vector that has m components equalling "1." Here, n is a quantity of neurons in the field and m is a quantity of neurons in a neuron ensemble; moreover, $m << n$. We call such binary vectors input masks. In order to feed a mask to the input of the associative field, it is necessary to excite the neurons of this field, whose numbers coincide with the numbers of the positions of unit elements in the binary vector mask.

The procedure for obtaining the pseudorandom k-bit numbers was developed on neurocomputer [1]. The pseudorandom binary vectors (masks) must possess the property that the probability of the appearance of 1 in any position has a predetermined value. The neurocomputer realization is based on the fact that the generation of such vectors is connected by the logical bit operations with the binary vectors.

The procedure is realized as follows. Let the number p, which indicates the probability of the appearance of a 1 in any position, be represented in the binary form, where $p_1, p_2, \ldots, p_k$ are the bits of this number and p_1 is the most significant bit. Furthermore, there is the uniformly distributed random binary number X, i.e., the probability of the appearance of a 1 is equal to 1/2. This vector is easily formed with a random-number generator. The required binary vector Y, i.e., the vector with the predetermined probability of the appearance of a 1, can be obtained as follows:

if $p_1 = 0$, then

$$Y = (P_2 \; \& \; (X_{21} \; \& \; X_{22})) \; U \ldots \; U(P_k \; \& \; (X_{kl} \; \& \; \ldots \; \& \; X_{kk})), \tag{5.8}$$

where P_2 is the vector that consists of all 1s if $p_2 = 1$, or of all 0s if $p_2 = 0$; P_k is the analogous vector determined by component p_k; all vectors X $(X_{21}, X_{22}, \ldots)$ with indices are independent pseudorandom vectors with the probability of 1s equal to 1/2: if $p_l = 1$, then the inversion of the number p is taken, on which the vector Y is formed, and it is inverted. With this approach, the maximum error does not exceed 12.5%.

Other procedures for the formation of vectors with the given probability are proposed, which have greater accuracy [1]. These procedures are used in the initial stage of the work of neural networks. The formed vectors (masks) are stored in the neurocomputer's memory. If it is necessary to feed the object that contains several features to the network input, then the object mask is formed as the bitwise disjunction of the feature masks. In this case, a quantity of unit elements in the mask exceeds m, and in order to preserve the approximately constant size of all ensembles, the normalization of the obtained object mask is produced. The normalization consists of the removal of excessive unit elements with the aid of the special procedure, which guarantees the approximately equal representation of all features in the resulting object mask. (The normalization will be described in detail later.)

One of the basic special features of associative-projective neural networks is the fact that any information can be represented as a neuron subset. Therefore, it is necessary to have a method for transforming different types of data into the subsets of neurons. Moreover, it is necessary to find such methods of mapping that the properties of the data are presented in some manner in the properties of neuron subsets. A certain metric could be one such property. If each set of the parameters corresponds to a point in the metric space, then it is possible to determine the distance between two sets. In order to preserve this property, it is necessary to determine the metric in the coding subsets and to ensure the similarity of the metrics. It is possible to consider a quantity of common unit elements in both subsets (overlap) as a natural measure of the proximity of two subsets. We will use this measure in this chapter for evaluating the codes obtained in this and another method of coding.

5.2.1.1 Coding the numbers and sets of the numerical parameters

To code the numbers, it is necessary to represent them in the form of discrete sets. In our work, we bring the numbers to the specific range and consider that they have only integer values (for example, in the range of 0 to 100). To code the numbers in this range, it is necessary for each number to generate the mask in such a manner that the masks with large overlap correspond to close numbers and the masks with small overlap correspond to distant numbers. Let us describe one of the procedures of mask creation that possesses this property and was used in different applied systems based on neural networks.

Let us generate the random binary vector with the probability of the appearance of 1s equal to m / n, where m is the size of the neural ensemble and n is a quantity of neurons in the neural field. With this selection of probability, a quantity of 1s in the vector will approximately equal m. We will consider the selected vector as the mask, which codes the number "0" (M_0). To obtain the mask that codes the number "1," let us build a new random vector X with the probability of the units of $p(X)$. Let us make a bitwise disjunction of vector X with the mask of the number "0":

$$Y = M_0 \cup X; \tag{5.9}$$

thereafter, we normalize the vector Y in such a way that it would also contain approximately m "1"s. The normalized vector Y is the mask that codes the number "1." Let us denote it as M_1.

It is not difficult to see that the mask that codes "1" will have a certain quantity of common unit elements with the mask that codes "0." This quantity decreases with an increase of the probability of $p(X)$. The mask that codes "2" is obtained from the mask that codes "1" with the aid of the same procedure. The random vector X is created anew, independently of the preceding one. The plot of the mask intersection versus the number values is shown in Fig. 5.8. Along the X-axis on the plot is given the difference of two number values, and along the Y-axis is given the quantity of common elements in their masks. The form of the curve represented in Fig. 5.8 plays a large role in the use of the associative-projective structures for the solution of pattern recognition problems.

Many objects in the environment are described not only by the presence or the absence of any features but also by the sets of numerical parameters. In order to code the set of numerical parameters, it is necessary to bring each of these parameters within the range selected for the masks. After this, it is necessary to create

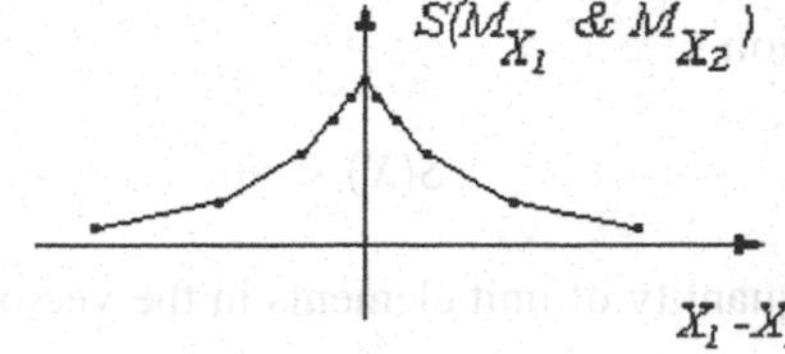

Fig. 5.8 The mask intersection

the masks that code each of the parameters (name of the parameters), and then to make the conjunction with the mask that codes the numerical value of this parameter. The conjunctive terms of the parameters obtained are united disjunctively for the parameter set, forming the nonnormalized code of the set of numerical parameters. Examples of such coding will be given in the description of concrete tasks.

A quantity of 1s in the coding masks must be selected in such a way that the total quantity of ones in the nonnormalized code of the parameter set exceeds the sizes of the ensembles, since all operations of normalization assume a decrease in the total quantity of 1s in the codes.

5.2.1.2 Code normalization

Codes are normalized in order to obtain the necessary sizes of the ensembles formed in the associative fields of higher levels. We will examine the normalization procedures of the vectors that code the feature sets.

Normalization procedures have to meet several requirements. One basic requirement consists of equal representation of any feature in the normalized code. This means that the normalized code has to contain approximately equal numbers of “1” drawn from the codes of the features that were combined. The second requirement states that during normalization of the codes for different feature sets, the representatives of the same feature must be different, i.e., the same representatives must not always be selected. The third requirement is that during repeated codings of the same feature set, the normalization must preserve the same representatives of each feature.

Let X be the binary vector subject to normalization and m be the quantity of unit elements that must remain in the vector after normalization. Let us select the set of the entire random numbers $s[1], s[2], \ldots, s[k]$, uniformly distributed in the interval $(0, n)$, where n is a quantity of neurons in the buffer field, and let us treat this set as the characteristic of the concrete buffer field in which the normalization procedure is produced. Let us denote as $Y(s)$ the vector Y cyclically shifted on s digits. Then the normalization procedure will appear as follows:

1. Form the working vector $Y = \bar{X}$, where $\bar{X}$ is the bitwise inversion of vector X, and $i = 1$.
2. Perform the operation

$$X = X \mathbin{\&} Y(s[i]). \tag{5.10}$$

3. Check the condition

$$S(X) < m, \tag{5.11}$$

where $S(X)$ is a quantity of unit elements in the vector X.

4. If the condition (4.11) is true, then the procedure is complete; otherwise, go to step 5.

5. $i = i + 1$. Go to step 2.

This procedure satisfies all requirements of feature coding. In the initial codes of the features, the unit element can appear with the probability determined during the mask formation. Since the feature masks are formed independently of each other, all the features in the assembled nonnormalized code have an approximately equal number of representatives (unit elements). The vector Y produced in step 1 of the procedure is the inversion of vector X; therefore, it has "0s" in the places where X has "1s". Each "0" in the shifted vector Y could eliminate a "1" from vector X. The probability that this unit belongs to any of the initial features is constant; therefore, a quantity of unit elements that will be preserved by different features should be approximately identical for all features. Consequently, the first requirement is fulfilled.

Note that different Y vectors produce partial extinction of unit elements. Thus, the feature that falls in different feature combinations loses different unit elements, fulfilling the second requirement. It is also obvious that during repeated coding of the same set of features, the same representatives remain in each feature because vector Y contains the same initial components, and constants of the shift of this vector are also the same (they represent the characteristic of the corresponding buffer field). It is possible to see also that during coding of the similar feature sets, many common representatives remain in each feature because the vectors Y are strongly correlated for similar sets. This leads to the large correlation of elements removed from the code of each feature, and therefore to the correlation of the remaining representatives of each feature.

It is not difficult to build the normalization procedures taking into account the priorities of different features on the basis of the described procedure. It is possible to form the feature set while fulfilling the normalization procedure, joining some features after this procedure is partially executed. Such features will preserve more representatives than those that were located in the set from the very beginning. It is possible to build the procedures of the formation and normalization of the feature set in such a way that the representatives of features will enter into the normalized set with practically any given weights.

Local connected coding was used to solve pattern recognition problems. The results of applying this coding will be described. However, local connected coding has specific deficiencies. One drawback of local connected coding during image recognition is that the identical elements of patterns located in different image places are represented by different stochastic codes, practically not connected with each other. This prevents the formation of the neural ensembles that present the separate elements of patterns and makes pattern recognition difficult when the object in the image lies in different positions. To eliminate these deficiencies, a new mechanism of neural coding is proposed. In this method, the name of the parameter is represented by the binary stochastic code, and its numerical value is represented by the shift of this code. We will identify this coding as shift coding.

5.2.2 *Shift Coding*

Shift coding was for the first time described in [8]. In shift coding, just as in local connected coding, the neural code of the feature name is the random binary vector. This code is stored in the memory of the neurocomputer and is extracted each time it is necessary to code or decode the feature. We term this vector the feature mask. The position of the feature in the image is coded by the appropriate shift of the feature mask. Let us consider the procedures of the shift in detail. We represent the code vector with a straight line (Fig. 5.9a) and the unit elements of this code vector with a "+."

Let us consider the coding procedure of a feature that can be formed along the coordinate axis *X*. On the line depicting the feature mask, we select several equidistant points shown in Fig. 5.9 by the vertical lines, and we term these points as points of absorption.

Let $X = 1$. In order to code this value, let us shift the entire mask one position to the right. If some unit elements fall into the points of absorption, they disappear, i.e., they are converted into the 0 elements (Fig. 5.9b). If, for example, it is necessary to code the value $X = 4$, then the right shift is carried out to four elements, and all unit elements that fall into the points of absorption disappear (Fig. 5.9c). The negative value of *X* is coded by shifting the mask to the left with the same disappearance of unit elements at the points of absorption. The value $X = -3$ is represented in Fig. 5.9d.

Let us consider another coding procedure of a feature moved along the coordinate *X* in which, instead of points of absorption, we use points of the cyclic closing of the shift code (Fig. 5.10a).

During shift coding with cyclic transfering, the elements falling into the points of transfer are moved to the opposite end of the corresponding segment. The value $X = +4$ will be represented by the code (Fig. 5.10b). While coding negative values, the transfer is accomplished in the opposite direction, for example, value $X = -2$ is shown in Fig. 5.10c. Shift coding with cyclic transfering can be used in the polar

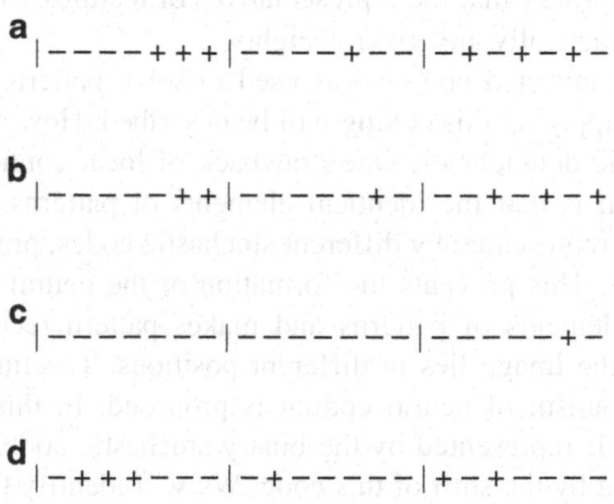

Fig. 5.9 Examples of codes

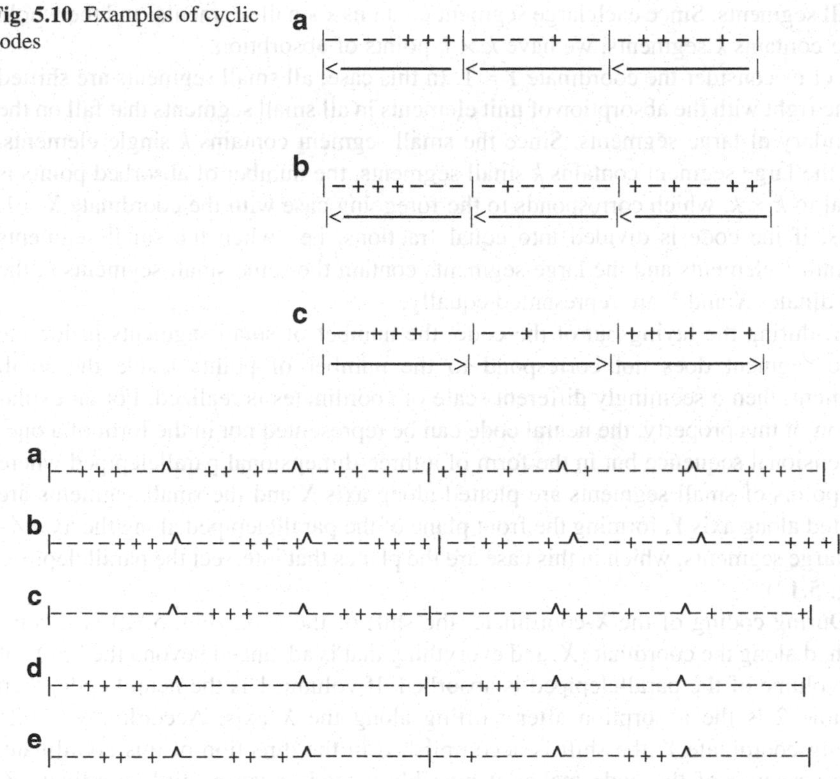

Fig. 5.10 Examples of cyclic codes

Fig. 5.11 Examples of shifting codes

coordinates, for example, for coding the angles of pattern rotation. For Cartesian coordinates, it is more convenient to use shift coding with absorption.

Let us consider the coding procedure for two coordinates *X* and *Y*. We shall break the line segments obtained after partition for coding coordinate *X* into smaller segments (Fig. 5.11a, where the ends of small segments are denoted by ^).

The coordinate *X* is coded just as in the preceding case, but the points of absorption are now the ends of the small segments. Thus, for example, in order to code the coordinates $X = +2$, $Y = 0$, it is necessary to shift the code two positions to the right with the absorption at all points marked in the figure by ^ (Fig. 5.11b). A change of the coordinate *Y* is coded not by bitwise shifts but by shifts to the size of the small segment. In this case, the boundaries of large segments serve as the points of absorption. Thus, Fig. 5.11c depicts the code of the coordinates $X = 0$, $Y = +1$; Fig. 5.11d depicts the code of coordinates $X = 0$, $Y = -2$; and Fig. 5.11e depicts the code of the coordinates $X = +1$, $Y = -1$.

Let us consider the case of shift coding with absorption when the small segment consists of *k* points. Let a quantity of small segments *k* be equal to a quantity of large segments. If the coordinate $X = 1$, then the shifts are accomplished inside the small segments by one digit to the right and are absorbed on the boundaries of the small segments. Since each large segment contains k small segments, and the whole code contains k segments, we have $k \times k$ points of absorbtion.

Let us consider the coordinate $Y = 1$. In this case, all small segments are shifted to the right with the absorption of unit elements in all small segments that fall on the boundary of large segments. Since the small segment contains k single elements, and the large segment contains k small segments, the number of absorbed points is equal to $k \times k$, which corresponds to the foregoing case with the coordinate $X = 1$. Thus, if the code is divided into equal fractions, i.e., when the small segments contain k elements and the large segments contain the same small segments k, the coordinates X and Y are represented equally.

If, during the laying out of the code, the number of small segments inside the large segment does not correspond to the number of points inside the small segment, then a seemingly different scale of coordinates is realized. For an explanation of this property, the neural code can be represented not in the form of a one-dimensional sequence but in the form of a three-dimensional parallelepiped where the points of small segments are plotted along axis X and the small segments are plotted along axis Y, forming the front plane of the parallelepiped along the axis Z - the large segments, which in this case are the planes that intersect the parallelepiped (Fig. 5.12).

During coding of the X-coordinate, the shift of the code (Fig. 5.12) is accomplished along the coordinate X, and everything that is advanced beyond the limits of the volume of the parallelepiped is absorbed. If volume 1 is the neural code, then volume 2 is the absorption after shifting along the X axis. Accordingly, while coding coordinate Y, the shift is accomplished in the direction of this coordinate. This structure of the code makes it possible to code even a third coordinate Z; however, we reserve this possibility for other purposes.

Let us examine the behavior of the neuron code representing the object that is described by the feature set extracted from the image. Each feature is shifted correspondingly to its position. The object code is formed as the bitwise disjunction of the codes of all features. This method of neural code formation leads to the fact that with the shift of the entire object on the image, there is no need for new coding of the feature positions as in the code of local connected coding. In order to code the new position of the object, it is sufficient to move its corresponding code, which

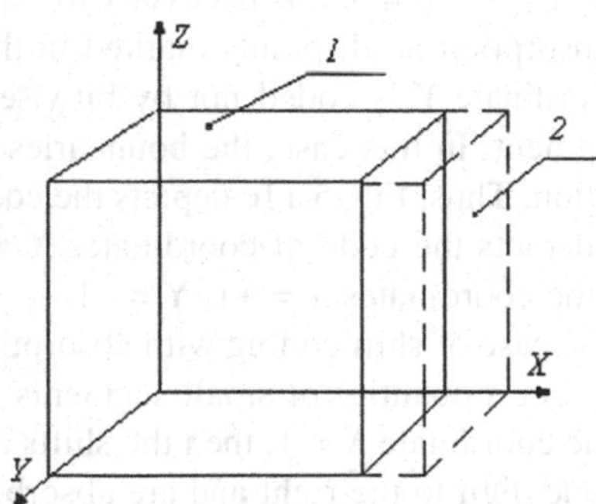

Fig. 5.12 Properties of neural codes

was formed in the other position of the object. We will use this property to train the neural network to recognize objects in the location of the arbitrary image.

Let a certain object (for example, a handwritten symbol) be located on an image of $W \times H$ pixels. To code its position on the image, it is necessary for the small segment to contain more than W points and for the large segment to contain more than H small segments. Keep in mind that the ensemble [17] corresponding to the neural code of an object in any position is formed in the neural network. Let us consider the algorithm that makes it possible to do such training. When the object is presented in the same position for the formation of a neural ensemble, we usually use the following rule of synaptic weight change:

$$w_{ij}^{r+l} = w_{ij}^{l} U(q_i \ \& \ q_j \ \& \ r_{ij}), \tag{5.12}$$

where w_{ij} is the binary value of synaptic weight before training (at the moment t) and after training (at the moment $t + 1$); q_i, q_j are the binary values of output signals of the neurons i and j; r_{ij} is the random binary variable; and U and & are the signs of disjunction and conjunction, respectively.

The neuron is active if the corresponding bit in the neural code equals "1." If the probability that $r_{ij} = 1$ equals $p(r_{ij})$, each pair of active neurons can obtain the exciting connection with this probability (at the start of training). If we have m active neurons, the total number of possible connections among them is m^2, and the mean number of connections really formed will be $p(r_{ij})^* \ m^2$. Such exciting connections form the neural ensemble. In order to form the neural ensemble that corresponds to the arbitrary position of the object, it is necessary to introduce an additional procedure into the training algorithm.

Let us examine an object on the image with coordinates X and Y. Let the object be moved on the image to the right into the point with coordinates $X + 1$, Y. Let us consider two neurons, i and j, which belong to the neural code of the object (Fig. 5.13a). After the shift, their positions are shown in Fig. 5.13b. According to expression (5.10), the connection for the initial image must be formed between the neurons i and j (the weight of the connection, which was "0," must become "1"). Let us consider the part of the matrix of synaptic weights around the intersection of line i and column j (Fig. 5.14a).

If the object is shifted one pixel to the right, the neurons i and j will be in the new positions $(i + 1)$, $(j + 1)$, and the additional unit value of the weight will appear in the matrix of synaptic weights at the intersection of line $(i + 1)$ and column $(j + 1)$ (Fig. 5.14b). If the object is moved more to the right, then the unit weight will appear at the intersection of the line $(i + 2)$ and the column $(j + 2)$, and so on. Subsequent shifts along coordinate X will lead to the component i being absorbed on the boundary of the small segment; therefore, no new weights in the matrix of synaptic weights will be formed (Fig. 5.14c).

a |– + – – –|...|+ – – – –|
i j

b |– – + – –|...|– + – – –|
i j

Fig. 5.13 Codes of object movement

a	*j*	**b**	*j*(*j*+1)	**c**	*j*
	0 0 0 0 0		0 0 0 0 0		0 0 0 0 0
i	1 0 0 0 0	*i*	1 0 0 0 0	*i*	1 0 0 0 0
	0 0 0 0 0	*i*+1	0 1 0 0 0		0 1 0 0 0
	0 0 0 0 0		0 0 0 0 0		0 0 1 0 0
	0 0 0 0 0		0 0 0 0 0		0 0 0 1 0

Fig. 5.14 Matrix of synaptic weights

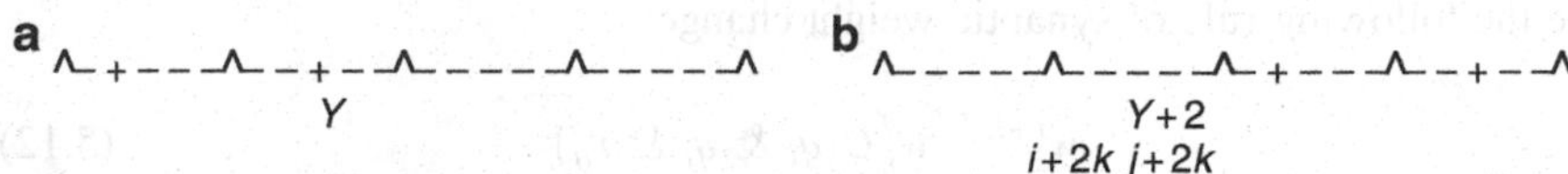

Fig. 5.15 Codes of object shifting

Let us consider the shift of the object along coordinate *Y*. In this case, the neural code is to be shifted by the given number of small segments. The code of the object located at the point with coordinates *X*, *Y* is presented in Fig. 5.15a. The code of the image at the point with coordinates *X*, *Y* + 2 will be as shown in Fig. 5.15b, where *k* is the number of points in the small segment.

For the initial position of the object, the synaptic weights are formed as represented in Fig. 5.16a. With the shift to the points with coordinates (*X*, *Y* + 1) and (*X*, *Y* + 2), the matrix of synaptic weights takes the form represented in Fig. 5.16b. The situation in which the shifts are carried out along the coordinates *X* and *Y* is shown in Fig. 5.16c.

In the case of cyclical permutation without absorption, the matrix of synaptic weights after the shifts both on *X* and on *Y* will take the form represented in Fig. 5.17. During this coding, the object or fragment corresponding to the formed ensemble will be recognized in any place of the image. It is obvious that the network's memory capacity in this mode of training will be less than that for local connected coding, since each new ensemble introduces considerably more unit elements into the matrix of synaptic weights (compared to each unit during local connected coding, $k*r/4$ unit elements appear here, where *k* is a quantity of points in the small segment, and *r* is a quantity of small segments in the large segment). Therefore, shift coding needs a neural network that considerably exceeds previous neural networks in size, i.e., the speed advantage of recognition obtained during shift coding requires high memory consumption. Since at present the cost of memory is small and is decreasing continuously, this approach makes sense.

5.2.2.1 Centering the shift code

For shift coding, it is necessary to know how to restore the value of the code and to determine the coordinates of the coded object. For this purpose, the concept of the

a

```
               j
    00000    00000    00000    00000
i   00000    00100    00000    00000
    00000    00000    00000    00000
    00000    00000    00000    00000
    00000    00000    00000    00000

    00000    00000    00000    00000
    00000    00000    00000    00000
    00000    00000    00000    00000
    00000    00000    00000    00000
    00000    00000    00000    00000

    00000    00000    00000    00000
    00000    00000    00000    00000
    00000    00000    00000    00000
    00000    00000    00000    00000
    00000    00000    00000    00000

    00000    00000    00000    00000
    00000    00000    00000    00000
    00000    00000    00000    00000
    00000    00000    00000    00000
    00000    00000    00000    00000
```

b

```
               j        j+k      j+2k
      00000    00000    00000    00000
i     00000    00100    00000    00000
      00000    00000    00000    00000
      00000    00000    00000    00000
      00000    00000    00000    00000

      00000    00000    00000    00000
i+k   00000    00000    00100    00000
      00000    00000    00000    00000
      00000    00000    00000    00000
      00000    00000    00000    00000

      00000    00000    00000    00000
i+2k  00000    00000    00000    00100
      00000    00000    00000    00000
      00000    00000    00000    00000
      00000    00000    00000    00000

      00000    00000    00000    00000
      00000    00000    00000    00000
      00000    00000    00000    00000
      00000    00000    00000    00000
      00000    00000    00000    00000
```

Fig. 5.16 (Continued)

c		*j*	*j+k*	*j+2k*
	00000	01000	00000	00000
i	00000	00100	00000	00000
	00000	00010	00000	00000
	00000	00001	00000	00000
	00000	00000	00000	00000
	00000	00000	01000	00000
i+k	00000	00000	00100	00000
	00000	00000	00010	00000
	00000	00000	00001	00000
	00000	00000	00000	00000
	00000	00000	00000	01000
i+2k	00000	00000	00000	00100
	00000	00000	00000	00010
	00000	00000	00000	00001
	00000	00000	00000	00000
	00000	00000	00000	00000
	00000	00000	00000	00000
	00000	00000	00000	00000
	00000	00000	00000	00000
	00000	00000	00000	00000

Fig. 5.16 Matrices of synaptic weights

shift code's center of gravity is introduced. Let us have a certain n-dimensional code containing, on average, m unit elements (for example, $n = 32{,}768$, $m = 1{,}024$). The center of gravity is determined from the formula

$$X_{cg} = \frac{\sum_{i=1}^{m} x_i}{m}, \quad Y_{cg} = \frac{\sum_{i=1}^{m} y_i}{m}, \tag{5.13}$$

where x_i is the coordinate of the unit element in the small segment, if we consider the beginning of each segment the origin of the coordinates; y_i is the coordinate of the small segment in the large segment, if we consider the beginning of each large segment as the origin of the coordinates; and m is the number of unit elements in the code. Summing is performed on all unit elements.

The following experiment was carried out. The random binary vector (code) with the values n, m assigned above was generated. The code's experimentally calculated center of gravity had coordinates X_{cg} = 16,86, Y_{cg} = 16,55. It was assumed that the object described by this code was moved to the point with the new coordinates (X, Y), and the code of the object was changed correspondingly.

	j	$j+k$	$j+2k$	
	00000	01000	00000	00000
i	00000	00100	00000	00000
	00000	00010	00000	00000
	00000	00001	00000	00000
	00000	10000	00000	00000
	00000	00000	01000	00000
$i+k$	00000	00000	00100	00000
	00000	00000	00010	00000
	00000	00000	00001	00000
	00000	00000	10000	00000
	00000	00000	00000	01000
$i+2k$	00000	00000	00000	00100
	00000	00000	00000	00010
	00000	00000	00000	00001
	00000	00000	00000	10000
	01000	00000	00000	00000
	00100	00000	00000	00000
	00010	00000	00000	00000
	00001	00000	00000	00000
	10000	00000	00000	00000

Fig. 5.17 Matrix of synaptic weights

After this, the code's center of gravity again was calculated. The results for several positions of the object are given in Table 5.1. Thus, the coordinates of the center of gravity of the code for the object located at the coordinates $X = -20$, $Y = 10$ are determined as $X_{cg} = 26{,}66$, $Y_{cg} = 11{,}26$.

To determine the coordinates of an object using the position of center of gravity, it is necessary to use the following formula:

$$X = 2 * (X_{cg} - X^{*}_{cg}), Y = 2 * (Y_{cg} - Y^{*}_{cg}), \tag{5.14}$$

where X_{cg}, Y_{cg} are the coordinates of the center of gravity of the initial code (object in the initial position), and X^{*}_{cg}, Y^{*}_{cg} are the coordinates of the center of gravity of the code in the new position (after shift). Using the tabular values, we will obtain $X = -19{,}6$, $Y = 10{,}58$, which represents the approximation of shift values.

In order to decode the name of the object, an additional associative field is introduced. In this field, the ensembles corresponding to the separate features and the names in the different positions are formed. The obtained code of the object is applied to the additional associative field, where its name and features in the appropriate position will be recognized. Then the codes of the recognized feature are centered and input to the decoder.

Table 5.1

Y	X									
	–20		–10		0		10		20	
–20	26,36	26,24	22,21	26,11	17,40	26,19	12,87	26,10	7,48	26,09
–10	26,48	21,79	21,79	21,48	16,97	21,46	12,80	21,26	7,71	21,21
0	26,59	17,03	21,72	16,65	16,86	16,55	12,73	16,43	7,68	16,43
10	26,66	11,26	21,41	11,38	16,47	11,43	12,62	11,56	7,75	11,57
20	26,48	6,46	21,34	6,54	16,58	6,42	12,50	6,40	7,54	6,25

5.2.2.2 Application of shift coding

The basic associative field of shift coding can be used for image recognition because it provides recognition independently of object position. Cyclic shift coding without absorption makes it possible to obtain the recognition invariant to the rotation of an object on the image, and only invariance to the scale of the image must be obtained by additional methods (for example, learning of the image simultaneously on several scales).

One positive quality of shift coding is that the smaller neural ensembles corresponding to the frequently occurring fragments of the image are formed automatically and simultaneously with the formation of the neural ensembles corresponding to the entire object. Thus, if we train this neural network for face recognition, the ensembles that correspond to eyes, nose, mouth, and so forth must be formed in the network. This quality can be used also for treating natural-language texts. If we code separate letters, and their positions in the word are coded by the corresponding shift, then the neural ensembles corresponding to the most frequent combinations of letters, including the roots of the words, the prefixes, and the suffixes, must be formed automatically. Similarly, at the training level of phrases, the ensembles corresponding to the most frequent word combinations must be formed. This property can be useful when creating information storage and retrieval systems. Local connected coding does not possess these properties, so it is necessary to use artificial methods to form analogous "small" ensembles.

The specific application of shift coding will be described in the following paragraphs since shift coding actually is the basis of multi-float coding. The classifiers built using multi-float coding showed sufficiently good results, which will be described below.

5.2.3 Functions of Neural Ensembles

The vectors formed by the method described above (for example, with the use of local connected coding) are used to train the associative field. Each unit in the vector corresponds to the active neuron, and 0 corresponds to the neuron that is not excited. During training, two active neurons are connected with binary connection.

If the connection exists, then its synaptic weight equals "1;" otherwise, it is "0." The associative-projective neural networks are stochastic networks, meaning that the connections between the active neurons are established with a certain probability. If the same vectors are input to the network, the probability of connections forming between the active neurons will increase. Thus, in the network are formed the sets of neurons having a higher probability of connection than the mean probability of connection in the whole network. We term such sets the neural ensembles [12].

In the ensemble, it is possible to distinguish the nucleus and fringe [13]. Neurons of the ensemble having a higher probablity of connection correspond to the nucleus. The most typical information about the presented object corresponds to the nucleus. The individual differences of the representatives of the object class correspond to the fringe. If we extract the different quantities of neurons with the greatest activity, for example, those assigning a high threshold of neural activity, then we can ensure a different level of concretization in the description of the object. For example, if the nucleus of the formed ensemble is named "apple," the extended description (taking into account the neurons entering the fringe of the ensemble) can contain the information "red, round, large." The description of the object of a different level makes it possible to speak about existence in the neural network hierarchy as "class – element of class," reflecting the subsumption relations.

The neural ensemble is the basic information element of all hierarchical levels of the neural network. It is formed from the elements of lower hierarchical levels and can correspond to the feature, to the description of an object, to the description of a situation, to the relation between the objects, and so forth. Its internal structure reflects the structure of the corresponding object. The fact that part of the ensemble is entirely excited allows us to consider the ensemble as a united and indivisible element in one hierarchical level. However, when it is transferred to other hierarchical levels, it is divided in such a way that only a part of its neurons is included into the descriptions of the more complex objects of upper hierarchical levels.

Assume, for example, that it is necessary to build the description of a tree, which consists of the stem, branches, and leaves. Each element has its own description. Thus, for example, leaves can have form, color, and texture. Let each of the named features in the associative field of the neural network's lower level be coded in the form of the subset of the neurons. Then, the neural ensembles corresponding to stem, leaves, and branches can be formed at the following hierarchical level. The neurons that describe its form, color, and texture at the lower level will enter into the ensemble that corresponds to the stem. So the sizes of the ensemble at the upper level will not be too large; only the part of the neurons from the ensembles of the lower level falls into the ensemble of the upper level. For example, during the construction of the ensemble that corresponds to the entire tree, only the parts of each of the ensembles describing the stem, branch, and leaves are included in it. We term the procedure for selecting the part of the neurons for transfer to the upper level the normalization of the neural ensemble. The ensemble is formed in such a way that, using the neurons that entered into the ensemble of the upper level, it would be possible to restore the ensembles of the lower level due to the associative reproduction of the entire ensemble from its part. This type of organization of the associative-projective neural network makes it possible to form the hierarchy as "part – whole."

The formation of neural ensembles is ensured by a change in the synaptic weights between the neurons of one associative field. Since the binary synaptic weights are used, the degree of an ensemble formation is characterized by the probability of unit synaptic weights between the neurons belonging to the ensemble. The better formed this ensemble, the higher the probability of establishing a unit connection between its neurons. Neural ensembles are formed in the associative field during training. Different training algorithms can be used. The Hebb training method works very well, as well as training with the delta rule (Widrow method), the Kohonen method (the process of self-organizing), the training law of Grossberg, and so on. In our case, we use Hebb's modified rule (Section 5.1.1).

5.2.4 *Methods of Economical Presentation of the Matrix of Synaptic Weights (Modular Structure)*

In many tasks of artificial intelligence, there appear problems of constructing associative neural networks of large dimensionality (containing a large quantity of neurons). Using fully connected neural networks leads to a quadratic increase of needed memory depending on the quantity of the neurons. With sufficiently large network dimensions (up to 10^6 neurons), the required memory becomes extremely large (up to 10^{12} bits). To construct networks of this volume, it is necessary to use structures that are not fully connected so that the needed memory can grow linearly with an increase of the quantity of neurons. We will consider two methods of constructing not fully connected neural networks: stochastic and modular. The modular method makes it possible to create a network with a constant ratio of the quantity of synapses to the quantity of neurons for large neural networks. To evaluate these different methods, we will use the following basic criteria: the convenience of hardware realization and the speed of the corresponding neurocomputer; the memory needed for neural networks; and the possibility of restoring the neural ensemble using its sufficiently small part.

5.2.4.1 Stochastic not fully connected networks

Before considering the stochastic method, let us recall that APNNs are characterized by the following features:

1. Any information elements (features, objects, relations, scenes, and so on) in such networks are presented by the neural ensembles.
2. The quantity of neurons entering the ensemble is much less than the total quantity of neurons in the network.

3. For any randomly chosen subset of neurons in the network, the following condition is satisfied: the power of the intersection of this subset with any neural ensemble does not depend on how this subset is selected; it depends only on its size and the sizes of the ensembles. The satisfaction of this condition is ensured by the method of coding information proposed in the works [8, 20] and described above. The primary features are coded with the subset of the neurons selected with the aid of a certain random procedure. The more complex ensembles are constructed from the "representatives" of primary features selected also with the aid of the special stochastic procedures. This leads to the uniform distribution of the probabilities of any neuron entering into any ensemble.
4. The synaptic weights can take only two values, "0" or "1."

If we let the network contain n neurons, then the fully connected network has the square matrix of synaptic weights containing $n \times n$ elements. The first method of constructing a not fully connected network consists of selecting the k elements, in which the formation of the connections ($k << n$) is permitted, from each matrix row with the aid of a certain random procedure. Computer memory is important only for these k element strings, and the procedure for selecting permissible elements is constructed in such a way that it would be possible to restore the position of these elements in each matrix row without high expenditure of memory. The corresponding procedure appears as follows. Let us consider the generator of pseudorandom numbers with a fixed initial state. As a result of this generator's work, let the sequence U ($u_1, u_2, \ldots, u_n$), which consists of the k-bit binary words, be generated. Let r be the relation n / k, and assume that r can be expressed in the form $r = 2^s$, where s is an integer.

Let us select the first s random numbers from the sequence U. From these numbers, let us form r binary numbers $p_1, \ldots, p_r$ according to the following rule:

$$\begin{aligned}
p_1 &= u_1 \ \& \ldots \& \ \bar{u}_{s-1} \ \& \ u_s, \\
p_2 &= u_1 \ \& \ldots \& \ u_{s-1} \ \& \ \bar{u}_s, \\
p_3 &= u_1 \ \& \ldots \& \ \bar{u}_{s-1} \ \& \ u_s, \\
p_4 &= u_1 \ \& \ldots \& \ \bar{u}_{s-1} \ \& \ \bar{u}_s, \ldots \\
p_{r-1} &= \bar{u}_1 \ \& \ldots \& \ \bar{u}_{s-1} \ \& \ u_s, \\
p_r &= \bar{u}_1 \ \& \ldots \& \ \bar{u}_{s-1} \ \& \ \bar{u}_s,
\end{aligned} \tag{5.15}$$

where $\overline{u_i}$ is the negation of u_i. It is not difficult to prove that any bit in the totality of the numbers $p_1, \ldots, p_r$ takes its unit value from one and only one of these numbers. Hence, this entire set of numbers contains exactly k unit bits. Conversely, if we carry out the bitwise disjunction of all numbers of this set, then the result contains unit elements in all bits. This circumstance can be used for the record of permissible connections in the first matrix row. If in this line we consecutively arrange the numbers $p_1, \ldots, p_r$, then all n bits will be filled ($n = r \times k$). In this case, unit bits will correspond to those row elements in which the establishment of interneuronal connections is permitted. An analogous operation is repeated for the second and subsequent matrix rows with new random numbers from the sequence U. In this case, the entire matrix of $n \times n$ connections will take in the memory n of k-bit words (one k-bit word for each row).

Let us consider an example. Let $n = 8$, $k = 2$, $r = 4$, $s = 2$ (Table 5.2). Let us select the sequence of random numbers $U = \{10, 10, 01, 00, 01, 00, 11, 01\}$. Let us describe the values p for the first matrix row,

$p_1 = 10 \ \& \ 10 = 10,$
$p_2 = 10 \ \& \ 01 = 00,$
$p_3 = 01 \ \& \ 10 = 00,$
$p_4 = 01 \ \& \ 01 = 01,$

and then enter them in the first matrix row. After performing analogous operations for all lines, we will obtain the matrix shown in Table 5.2. Unit elements in the matrix correspond to permitted interneuronal connections. After training, the weights of interneuronal connections will appear on the permitted positions as shown in Table 5.3. To store these weights in memory, they are collected from each matrix row into the compact word. The resulting compact array is shown in Table 5.4. While the network calculates neural activity, the conversion of the compact array into the complete matrix is accomplished, as shown in Table 5.3.

Advantages of this method are the compact presentation of the not fully connected matrix in the memory of the computer and the uniform distribution of the permitted connections over the matrix. The disadvantage is the high expenditure of time for the conversion of a compact array into a complete matrix. This conversion procedure is comparatively complicated to execute in parallel with the aid of the hardware; therefore, it is expedient to find a method of constructing not fully connected networks that would allow compact storage of information about the connections but that would not require such a complex procedure for restoring

Table 5.2

10	00	00	01
00	01	00	10
00	01	00	10
01	10	00	00
00	10	00	01
01	00	10	00
00	01	00	10
00	10	00	01

Table 5.3

W_{11}	0	0	0	0	0	0	W_{18}
0	0	0	W_{24}	0	0	W_{27}	0
0	0	0	W_{34}	0	0	W_{37}	0
0	W_{42}	W_{43}	0	0	0	0	0
0	0	W_{53}	0	0	0	0	W_{58}
0	W_{62}	0	0	W_{65}	0	0	0
0	0	0	W_{73}	0	0	W_{77}	0
0	0	W_{83}	0	0	0	0	W_{88}

Table 5.4

W_{11}	W_{18}
W_{24}	W_{27}
W_{34}	W_{37}
W_{42}	W_{43}
W_{53}	W_{58}
W_{62}	W_{65}
W_{73}	W_{77}
W_{83}	W_{88}

Table 5.5

block 1	block 2	block 3	block 4
xx xx $\overline{\text{xx}}$ xx	xx xx xx xx	xx xx xx xx	xx xx xx xx
module M_{13}			

Table 5.6

M_{11}	M_{12}	M_{13}	M_{14}^{*}
M_{21}^{*}	M_{22}	M_{23}	M_{24}
M_{31}	M_{32}	M_{33}	M_{34}
M_{41}	M_{42}	M_{43}^{*}	M_4

the complete matrix. Below, we will consider the method developed and utilized by us for neural networks having a large quantity of neurons [7].

5.2.4.2 Constructing modular neural networks

To construct a modular neural network, the entire set of n neurons is divided into k blocks, and each of these blocks is divided on k modules. Table 5.5 gives an example of the partition when $k = 4$. (In Table 5.5, x is the position of a neuron). Let us designate the modules through M_{ij} (i is the number of the block, j is the number of the module inside the block) and place them in the rectangular matrix (see Table 5.6). The modules of one block are located in the matrix row. Let us connect the outputs of the neurons belonging to the modules of the first matrix column with the inputs of the neurons belonging to the modules of the first matrix row. We say that module M_{ij} is connected with module M_{rs} if each neuron from M_{ij} is connected with all neurons from M_{rs}.

Let us connect the outputs of the second column with the inputs of the second row, and so forth. We will call this neural network structure the modular structure.

Let us consider some of its properties. It is clear that each block contains n / k neurons, and each module contains $n / (k \times k)$ neurons. In accordance with connection rules, the output of each neuron is connected with the inputs of all neurons of one block. Thus, each neuron has n / k connections. The total quantity of connections will be respectively equal $(n \times n) / k$. If increasing the number of neurons n proportionally increases a quantity of blocks k, then the total quantity of connections will grow linearly depending on the number of neurons. The limitation is that when no more than one neuron remains in each module, further increasing the quantity of neurons in the network no longer can be compensated by increasing the quantity of blocks k.

Although the modular structure of the network does not ensure the connection of each neuron with other neurons, it is possible to connect one neuron with any other through not more than one intermediate neuron. To prove this fact, let us consider two neurons, one belonging to module M_{ij} and the other to module M_{pq}. The neuron from the first module can be connected with the neuron from the second module through the intermediate neurons in module M_{jp}. Actually, the outputs of module M_{ij} are connected with all neurons of block j and therefore with module M_{jp}. The outputs of module M_{jp} are connected with all neurons of block p and therefore with the module M_{pq}, so for each neuron pair from modules M_{ij} and M_{pq}, we can find the neuron from module M_{jp} that connects this pair. For example, it is necessary to connect the neurons of module M_{21} with the neurons of module M_{43} (they are marked * in Table 5.6). The neurons of module M_{21} are connected with the inputs of block 1 and therefore with module M_{14}, whose outputs are connected with the inputs of block 4 and therefore with the neurons of module M_{43}.

The modular structure of the neural network makes it possible to create simple and highly effective hardware for its realization and to obtain neural networks that possess sufficiently high storage capacity. Storage capacity is a quantity of statistically independent neural ensembles that can be formed in the network and restored from their parts.

5.3 Conclusion

At present, there is a great variety of operation algorithms and architectures of neural networks that serve as a base for artificial intelligence systems. To solve the broad class of artificial intelligence problems, it is necessary to construct neural networks taking into account the principles proposed below.

1. Neural networks are to be constructed from standard functional blocks. For artificial intelligence systems, it is necessary to use neural networks that have a large quantity of neurons (no less than several thousand); therefore, the basic elements for neural network creation must be large functional blocks that have an internal organization and properties known to the developer. In our case, we propose using associate and buffer neural fields as the functional blocks.
2. Neural networks must have hierarchical organization. Many tasks of artificial intelligence cannot be solved without hierarchical structures that make it possible to create models of complex objects from simpler ones.
3. Integrity and separability of information elements are the principles of neural network construction. The information element in each hierarchical level has to behave as a single object, but it has to allow splitting upon transfer from one level to another. Upon transfer from an upper to a lower hierarchical level, this splitting corresponds to the extraction of its components, and upon transfer from

a lower level to an upper level, it corresponds to the introduction of the specific part of this element to the more complex object.

4. The information element must join the object name and description. The name and description of the object must be presented in all information structures into which the model of the object enters. Actually, the name of the object must be one of the features present in the description of this object.

In this chapter, we described the approach that makes it possible to construct neural networks that satisfy the enumerated principles. At the basis of this approach lies the idea that any object must be presented not by a separate symbol or a separate neuron, but by a subset of neurons (neural ensemble). Information coding and conversion of the codes in the ensembles of different hierarchical levels are of great importance in this approach.

Two coding methods that can be used to present information in the associative-projective neural networks were examined. One of them (local connected coding) was tested during the solution of problems of pattern recognition, which will be described later. This method has many advantages; however, to obtain the invariance of the codes relative to the object position on the image, it is necessary to use the special methods, which generate additional problems. To overcome this disadvantage, the second coding method (shift coding) was developed. This method makes it possible not only to obtain the invariant to the shift object presentation but also to automatically form the neural ensembles that correspond to the components of the object. It seems to us that this property can be very useful both during the solution of pattern recognition problems and during the creation of adaptive information retrieval systems.

The problem of saving memory in which the synaptic weights of connections are stored appears with neural networks that have a large quantity of neurons. We solved this problem by partitioning neural networks into separate modules and introducing special procedures for connecting these modules. The result in this case is a linear (but not quadratic) increase in memory requirements, as a function of the number of neurons.

References

1. Kussul E.M. Associative Neural Structure. Kiev: Nauk. Dumka, 1992, p. 144 (in Russian).
2. Amosov N.M., Baidyk T.N., Goltsev A.D. et al., Neurocomputers and Intelligent Robots, Ed. Amosov N.M., Kiev: Nauk. Dumka,1991, p. 272 (in Russian).
3. Baidyk T.N. Information Capacity Investigation of the Associative Field with Ensembles of Complex Structure, *Neural networks and neurocomputers*, Kiev, Institute of Cybernetics, 1992, pp. 39–48 (in Russian).
4. Kussul' E.M., Baydyk T.N. Design of a Neural-Like Network Architecture for Recognition of Object Shapes in Images. *Soviet Journal of Automation and Information Sciences (formerly Soviet Automatic Control)*, Vol. 23, No. 5, 1990, pp. 53–59.
5. Kussul E.M., Baidyk T.N. Some Functions of Associative Neural Field, Kiev, 1992, p. 21 (Preprint / Institute of Cybernetics; 92–5) (in Russian).

6. Kussul E.M., Baidyk T.N. Neural Assembly Structure, *Neurocomputer*, 1992, No. 1, pp. 16–23 (in Russian).
7. Kussul E.M., Baidyk T.N. Module Structure of Associative Neural Networks, Kiev, 1993, p. 16 (Preprint / Institute of Cybernetics; 93–6) (in Russian).
8. Kussul E.M., Baidyk T.N. Information Codification in Associative-Projective Neural Networks, Kiev, 1993, p. 18 (Preprint / Institute of Cybernetics; 93–3) (in Russian).
9. Kussul E.M., Rachkovskij D.A., Baidyk T.N. Associative-Projective Neural Networks: Architecture, Implementation, Applications, Proc. of Fourth Intern. Conf. "Neural Networks & their Applications", Nimes, France, 1991, pp. 463–476.
10. T.N.Baidyk, 2001, Neural Networks and Artificial Intelligence Problems, Kiev, Naukova Dumka, p. 264 (in Russian).
11. Lansner A., Ekeberg O. Reliability and Speed of Recall in an Associative Network, *IEEE Transactions on Pattern Analysis and Machine Intelligence*, PAMI–7, No. 4, 1985, pp. 490–498.
12. Hebb D.O. The Organization of Behaviour, New York: Wiley, 1949, p. 319.
13. Milner P.M. The Cell Assembly: Mark 2, *Psychol. Rev.*, Vol. 64, No. 4, 1954, pp. 242–252.
14. Kohonen T. Self-organization and Associative Memory, Berlin, Springer, 1984, p. 255.
15. Jackel L.D., Howard R.E., Denker J.S. et al. Building a Hierarchy with Neural Networks: an Example - Image Vector Quantization, *Appl. Optics*, 1987, Vol. 26, No. 23, pp. 5031–5034.
16. Nakano K. Associatron - a Model of Associative Memory, *IEEE Transactions on Systems, Man, and Cybernetics*, SMC–2, No. 3 July 1972, pp. 380–388.
17. Kussul E., Baidyk T. Structure of Neural Assembly, The RNNS/IEEE Symposium on Neuroinformatics and Neurocomputers, Rostov-on-Don, Russia, 1992, pp. 423–434.
18. Kussul E.M., Rachkovskij D.A., Baidyk T.N. On Image Texture Recognition by Associative-projective Neurocomputer, Proc. of the Conf. "Intelligent engineering systems through artificial neural networks", Ed. C. H. Dagli, S. Kumara and Y. C. Shin, ASME Press, 1991, pp. 453–458.
19. Tsodyks M.V. Associative Memory in Asymmetric Diluted Network with Low Level of Activity, Europhys. Lett. 7 (3), 1988, pp. 203–208.
20. Artikutsa S.Ya, Baidyk T.N., Kussul E.M., Rachkovskij D.A., Texture Recognition with the Neurocomputer, Kiev, 1991, p.20 (Prepr. / Institute of Cybernetics; 91–8) (in Russian).

Chapter 6
Recognition of Textures, Object Shapes, and Handwritten Words

Neural networks are widely used for solving pattern recognition problems [1–5]. Let us examine the following stages in the recognition of visual patterns: extraction of textural features; recognition of textures; extraction on the image of the regions with uniform texture; and the recognition of the shape of these regions. All these stages can be realized effectively on neurocomputers.

6.1 Recognition of Textures

Some statistical characteristics of the local sections of images usually are understood by textural features [6, 7]. We understand under the texture a property of the local section of the image, which is fixed in a certain extensive section of the image. The foliage of the trees, grass, asphalt coating, and so on can serve as examples of sections of images having identical texture.

6.1.1 Extraction of Texture Features

Let us assume that it is necessary to extract texture features in the local section of the image. We limit this local section with the rectangular window having dimensions $b \times h$ pixels^2. We will assume that in this window the following texture features are extracted: the histograms of brightnesses, of contrasts, and of orientation of contour elements.

The histogram of brightnesses is calculated as follows: the entire range of possible brightnesses is divided into k intervals. Thereafter, a quantity of pixels whose brightness falls into this interval is calculated for each interval. The number obtained for each interval is considered the parameter, which characterizes the

E. Kussul et al., *Neural Networks and Micromechanics*,
DOI 10.1007/978-3-642-02535-8_6, © Springer-Verlag Berlin Heidelberg 2010

texture in this window. The histogram of brightnesses gives as many texture parameters as the number of intervals into which the entire range of brightness was divided. Sometimes it is useful to add to these parameters an estimate of the mathematical expectation and dispersion of brightnesses in the window.

The histogram of contrasts is similar to the histogram of brightnesses, only instead of brightness, the difference in the brightnesses between adjacent pixels must be taken. In the histogram of the orientations of contour elements, the role of brightness plays the orientation angle of every contour element. The latter is defined as a vector, perpendicular to the gradient of brightness, determined by four adjacent pixels. The algorithm that helps us determine the orientation angle is developed using the Schwarz method [8]. The orientation angle of the contour element is considered in the histogram only when the value of the gradient exceeds a certain threshold.

In general, few textures have only a small number of sufficiently clear distinguishing features. Most frequently, it is necessary to take into account the complex combinations of a large quantity of features during texture recognition. In our case, each component of each histogram is considered a separate textural feature. Depending on the quantity of intervals in each histogram, the total quantity of textural features can vary from several to several hundred. This feature space can seem too large, but if we want to obtain a sufficiently universal system of technical vision, we must take into account the fact that, a priori, we do not know what features will be more important during the recognition of concrete textures. It is necessary to create a recognition system in which the excess of parameters does not interfere with recognition and does not greatly decrease the operating speed.

To enumerate all the features described above, it is possible to propose very simple and inexpensive hardware, which will perform this procedure in real time. We will consider that such hardware exists in the neurocomputer system of technical vision.

6.1.2 The Coding of Texture Features

The code of the texture sample is assembled from the obtained codes of feature names and the codes of the numerical values of these features as follows:

$$Z = \overset{k}{\underset{i=1}{U}} (M_i^{np} \mathbin{\&} M_i^j) \tag{6.1}$$

where k is the number of features; M_i^{np} is the mask of the feature name; and M_i^j is the mask of the number corresponding to the numerical value (j) of the feature i. The vector Z is supplied for the training of the associative-projective neural network.

6.1.3 Texture Recognition

The task of texture recognition has characteristic properties. On the one hand, it is impossible to extract a small number of informative features that would sufficiently describe the textures, and therefore it is necessary to deal with a large feature space. On the other hand, even a comparatively small quantity of images gives a large number of feature samples because the dimensions of the texture window are many times less than the dimensions of the image, and each image can give thousands of samples of textures during training. This large size of the training set makes it possible, even in multidimensional space, to build good dividing surfaces for classification.

Let us examine how texture recognition occurs with the aid of the neurocomputer. The system block diagram intended for texture analysis is represented in Fig. 6.1. The system consists of a TV camera T, a unit of the feature extraction FE, a coding block CU_1, a buffer neuron field BF, an associative neuron field AF, a decoding block CU_2, and a control computer MC. The blocks enclosed in the figure by the broken line are realized in the neurocomputer.

Let us consider the supervised training mode. Initially, the operator marks all the images that will be used for training. The operator indicates in the image the texture name for each possible position of the window, and this information is written in the memory of the computer. Thereafter, the first image from the training set is input into the block FE, and the computer puts the window in the initial position (for example, into the upper left corner of the image). The block FE begins the textural feature extraction within the limits of the window and transfers them into the coding block CU_1, from which the codes of features enter the buffer field BF. The buffer field BF creates the code of the texture using codes of the features.

The code obtained in the buffer field is normalized and transferred to the associative field AF, where neuron excitations during several cycles are calculated. Then, the output vector of field AF is transferred to the decoding block CU_2 to determine the code of the texture name with which it has the greatest intersection. This texture name is considered the recognized texture. It is compared with the name of the texture that the operator assigned for this position of the window. If the names coincide, then training is not required, and the computer moves the window to the following position (for example, to the right by half of the width of the window). Of course, in the beginning of training the recognized name of the texture

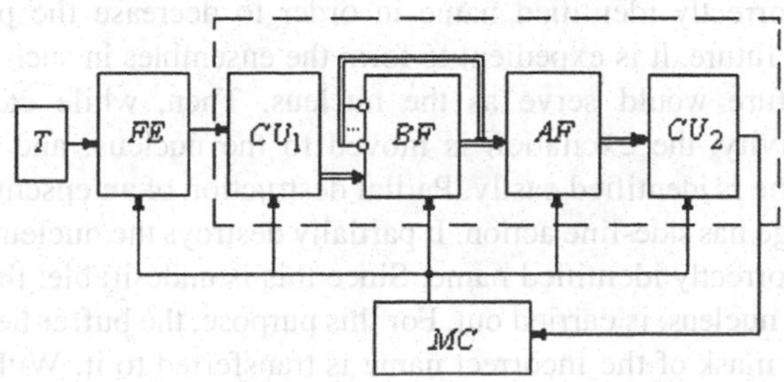

Fig. 6.1 Architecture for texture analysis

frequently will be incorrect, and then the procedure of training, which includes three stages, begins.

During the first stage, training with the correct name is carried out. For this purpose, the computer *MC* introduces the correct name of that texture into the coding block CU_1, which corresponds to the current position of the window. The coding device generates the mask of the correct name and transfers it into the buffer field *BF*, where it is united with the code of texture features and is transferred to create the corresponding ensemble in the associative field *AF*. The creation of the ensemble is achieved due to a change in the synaptic weights according to the formula:

$$W_{ij}{}' = W_{ij} \,\mathrm{U}\, \left(q_i \;\&\; q_j \& \; h_{ij}\right), \tag{6.2}$$

where $W_{ij}{}'$ is the synaptic weight of the associative connection from neuron *j* to neuron *i* after training; W_{ij} is the synaptic weight of this connection before training; q_i is the output of the neuron *i*; q_j is the output of the neuron *j*; h_{ij} is the binary random value with an average value that equals the signal of reinforcement ($M(h_{ij}) = l$); U is disjunction; and & is conjunction.

In the second stage, the ensemble with the incorrect name is partially destroyed. To complete this stage, the *computer* gives the command to repeat calculation, coding, and normalization of the texture features, after which it sends the name of the texture that erroneously was indicated by the recognition system into the coding block CU_1. The coding device finds the mask of this erroneous name and transfers it into the buffer field *BF*, at which point the code of the feature and the mask of the erroneous name are transferred to the associative field *AF*, where a change of the synaptic weights is produced with the negative signal of reinforcement according to the formula:

$$W'_{ij} = W_{ij} \& \overline{\left(q_i \;\&\; q_j \;\&\; h_{ij}\right)}, \tag{6.3}$$

where the horizontal line above the parentheses is negation and the remaining designations are the same as in formula (6.2), except that the mathematical expectation of the binary random variable h_{ij} is equal to the absolute value of the signal of reinforcement ($M(h_{ij}) = |l|$).

This operation is necessary because we need to decrease the total quantity of synaptic connections from the code of the analyzed texture to the mask of the incorrectly identified name in order to decrease the probability of this error in the future. It is expedient to form the ensembles in such a way that the name of the texture would serve as the nucleus. Then, while calculating associative field activity, the excitation is moved to the nucleus, and the corresponding texture name is identified easily. Partial destruction of an ensemble executed in the second stage has side-line action. It partially destroys the nucleus, which corresponds to the incorrectly identified name. Since this is undesirable, the third stage, resolution of the nucleus, is carried out. For this purpose, the buffer field, *BF*, is cleaned and only the mask of the incorrect name is transferred to it. Without the change, this mask enters the associative field, AF, where the formation of the ensemble is carried out with positive reinforcement according to formula (6.2). In this case, a large reinforcement value is selected, for example, $l = 1$.

After the fulfillment of all three stages, the computer transfers the window to the following position, and the procedure is repeated until the entire image is looked over.

6.1.4 The Experimental Investigation of the Texture Recognition System

The experimental study of the texture recognition process was carried out with photographs of a real environment. The investigations were conducted using a neurocomputer [9]. The photographic images taken on city streets and printed on photographic paper were introduced to the computer with the aid of a black-and-white scanner having 37 gradations of gray. The sizes of the images used in the experiment were 200×144 pixels^2. The size of the window was 16×16 pixels^2, and the step of the window displacement in horizontal and vertical directions was eight pixels. 408 window positions were examined in each image.

In the experiment, the following texture features were extracted:

1. Brightness features: 18 parameters in the histogram of brightnesses, average brightness, and the standard deviation of brightnesses within the window. 20 features in total.
2. Contrast features: 18 parameters in the histogram of contrasts, average contrast, and the standard deviation of contrast. 20 features in total.
3. The features of the contour element orientation: 32 parameters in the histogram of orientations, average orientation, and the standard deviation of orientations. 34 features in total.
4. The feature position (height) on the image. This feature is the vertical coordinate of the position of the window on the image. Although the position of different objects on the image is not fixed, there are some preferences. For example, the sky more frequently occupies the upper part of the image while asphalt coating tends to occupy the lower part. This feature, of course, is not texture. It is added as an element of a priori information. Generally speaking, a priori information in the associative-projective neural structures is introduced with the aid of the additional associative fields, which have hierarchical organization. However, in this experiment, a one-level network was used, and additional information about the preferential arrangement of textures on the image was introduced in the form of separate features.

Five classes of textures were used in the experiment: sky, trees (crown), road, transportation means, and posts or trunks of trees. The experiment was conducted as follows. First, the operator marked images, during which time the file indicating the texture name for each position of the window was formed on the computer. Then, this marked image was presented to the recognition system (408 different samples of textures). The training mode of the recognition system was switched off, and during the recognition process the number of errors made by the system was calculated. The error was calculated in any disagreement of the results obtained by the system with the marks of the operator, even when two or three different textures fell within the limits of the window. After this, the training mode was switched on, and the recognition system again examined the same image being trained for the recognition of its textures. Then the training mode was switched off, and the following image was presented to the system for the verification test, and so on.

Ten images were presented to the recognition system. Examples of the images are given in Fig. 6.2. Figure 6.3 gives the curve of the training of the system with the value of the reinforcement signal $l \approx 1/64$. The ordinal numbers of images are plotted along the X-axis while the percentages of the textures correctly identified before the training on this image are plotted along the Y-axis. In this figure, it is evident that the last three images were recognized with more than 70% probability. To increase the recognition rate, it is necessary to use additional color features and also to increase the number of images in the training sample. The described experiments were carried out via a method by which the extraction and coding of

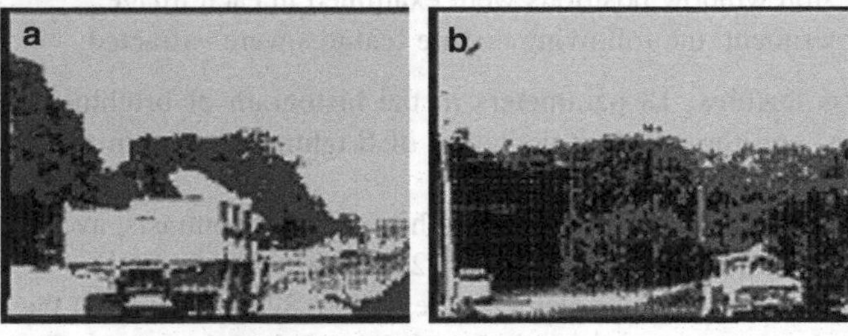

Fig. 6.2 Examples of real images

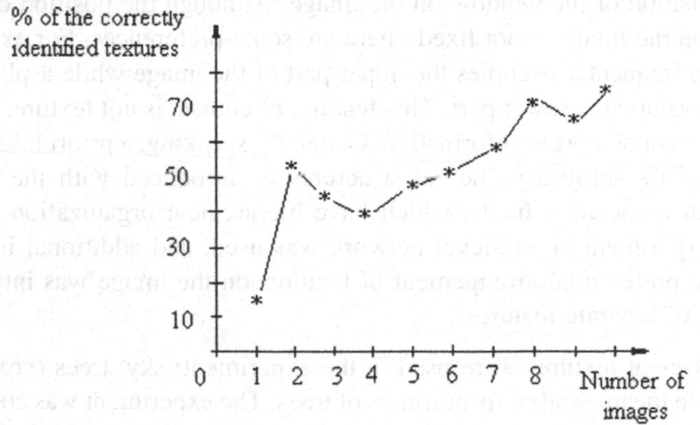

Fig. 6.3 Training of the system

texture features were accomplished on a personal computer while training of the neural network and recognition were accomplished with the neurocomputer.

The preliminary work for applying neural networks in tasks of stereopsis [9] was executed on the basis of texture recognition results based on the examples of the textures of Julesz [7]. Specialists in pattern recognition and systems of technical vision frequently focus their attention on tasks of stereovision and recognition of dynamic objects [6, 7, 10, 11]. However, because of the limited size of this book, this material is not presented here.

The problem of natural texture recognition was solved also by the method of potential functions. In order to estimate the results of texture recognition with the aid of both methods (neural networks and the method of potential functions), the method of potential functions was realized on a personal computer and tested on the same textures. Let us examine this method and the results obtained with it.

6.1.5 Texture Recognition with the Method of Potential Functions

The method of potential functions was developed for pattern recognition, and it has a good geometric interpretation [12]. The training process in the method of potential functions is related to the construction of the multidimensional surface that divides the feature space from the regions corresponding to different classes of patterns. The regions are not previously determined, and there is no information about the boundaries of these regions. The training points from these regions labelled with class names are presented to create dividing surfaces. The target of training is to build surfaces that will divide not only the presented points but also all the remaining points belonging to these regions. It is possible to create the functions $f_i(x)$ (x belongs to the feature space X). Each function $f_i(x)$ has to be positive in the region of the ith class and negative in other regions.

Functions similar to electrostatic potentials are constructed around each point of the training sample. The sign of the potential source depends on the class that corresponds to the selected point. Training in this case is connected with the points from the training sample being written into memory. Every point has a class indicator. During the recognition process for the new point, it is necessary to calculate total potentials, generated by all the points of each sign, and the point will correspond to the class with the maximum potential. We will term the points of the training sample saved in the memory support points. The potential functions are assigned in the form of simple decreasing functions. If we assign the distance of $d(x_i, x_j)$ between support point i and new point j, the potential function can be defined as $1 / (1 + ad)$, where a is the parameter of the slope of the potential function. (The more complex the dividing surface, the more rapidly the potential must diminish.) The parameter of the slope is usually selected experimentally.

Let us examine one of the examples of potential functions. Let us denote the training points belonging to class s through $X_i^s = (x_{i1}^s, \ldots, x_{ik}^s)$, where the index i indicates different support points and the letter k indicates a quantity of features in the feature space. Accordingly, let us denote the points belonging to class q as $X_j^q = (x_{j1}^q, \ldots, x_{jk}^q)$. Then the potential functions f_s and f_q for the classes s and q can take the following form:

$$
\begin{aligned}
f_s &= \sum_i \frac{1}{(x_1 - x_{i1}{}^s)^2 + \ldots + (x_k - x_{ik}{}^s)^2 + a}, \\
f_q &= \sum_j \frac{1}{(x_1 - x_{j1}{}^q)^2 + \ldots + (x_k - x_{jk}{}^q)^2 + a}
\end{aligned}
\tag{6.4}
$$

where $x_1, \ldots, x_k$ are the coordinates of the arbitrary point of the space, and a is the small positive value which prevents division into zero. If a quantity of classes is more than two, then similar functions $f_r, f_w \ldots$ are written for the remaining classes, and during the recognition the maximum $max(f_s, f_q, f_r, f_w, \ldots)$ is found.

In order to decrease the quantity of support points in the memory, only those points that cannot be recognized correctly using previously collected support points are saved. This procedure leads to the concentration of support points along the surfaces that divide the patterns, thus improving the quality of recognition. Nevertheless, in the spaces of large dimensionality with nonlinear dividing surfaces in the presence of the multiconnected regions, the quantity of support points must be very large. In the task of recognizing natural textures, the number of support points can reach several thousand. The method was programmed and realized for textural features described above. The comparative analysis of the results of the texture recognition is given below.

The method of potential functions was tested in ten photographs described above (five texture classes). The results of the texture recognition with the aid of APNN and the method of potential functions are given in Fig. 6.4. (1 – by the method of potential functions, 2 – with the aid of associative-projective neural networks). The percentage of correct recognition by the method of potential functions is somewhat higher than that with the aid of APNN. The number of support points in the method of potential functions achieved is 1,712.

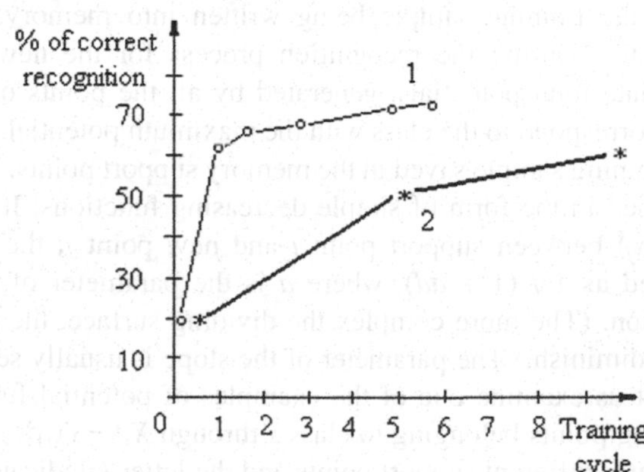

Fig. 6.4 Texture recognition with the aid of APNN (curve 2) and the method of potential functions (curve 1)

Now let us consider both the common properties of and the differences between the two approaches to texture recognition. If we consider as potential functions F a total quantity of synaptic connections from the neurons of input vector I to the neurons of the texture name T, then it is possible to form an analogy between the method of potential functions and the work of associative-projective neural networks. If the new point in the feature space is very close to one of the support points (i.e., a training sample that participated in forming the synaptic connections), then the mask of this support point will have much in common with the code of the new point (i.e., a quantity of connections of the input code with the class name is the analog of the potential field).

However, there are essential differences between recognition using APNN on a neurocomputer and recognition using the method of potential functions. The first difference is that on a neurocomputer, stochastic codes are used and potential functions $F(I, T)$ are random. The stochastic nature of these analogs of potential functions negatively influences the results of recognition if the quantity of neurons in the ensemble is only in the tens (the number we had on the neurocomputer). Increasing the number of neurons in the ensemble will decrease this influence.

The second difference relates to the speed of the neurocomputer and has special importance for the selection of the neurocomputer as a recognition device. Increasing the number of support points does not significantly increase the time of recognition on the neurocomputer, whereas in the method of potential functions, recognition time is directly dependent on the quantity of support points. In Fig. 6.5, the regions recognized by different methods in the same period (a – by the method of potential functions; b – by the method of the nearest neighbor; c – by APNN) are marked with dark colors. During the time that APNN spent to recognize the whole image, the method of potential functions recognized only a small part of the image (Fig. 6.5a).

6.2 Recognition of Object Shapes

We will not analyze in detail the existing approaches to the recognition of shapes. Let us note that one of the first methods of shape recognition was comparison with the image standards. This method is not satisfactory for several reasons. Firstly, the

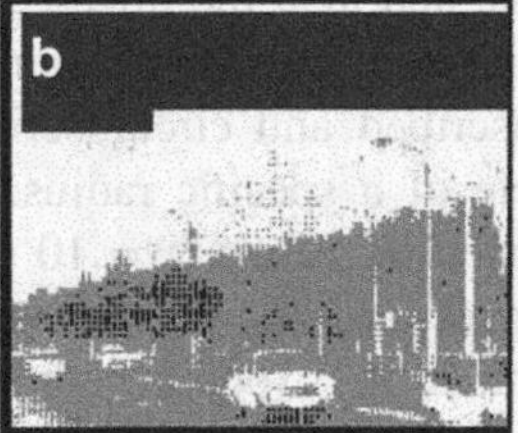

Fig. 6.5 Comparative time of different recognition methods. Colored parts correspond to the recognized area in the same time period (a-potential functions, b-nearest neighbor, c-APNN)

method does not work if we change the aspect angle for obtaining the image. Any rotation or change in the scale requires new standards, which in turn require large storage capacities. Secondly, a large quantity of standards leads to an increase in object recognition time. In order to recognize a shape that does not depend on affine transformations, researchers developed mathematical methods of converting the parametric spaces. One such method is the Hough transform [13, 14], though its application requires complex calculations. One of the promising trends in shape recognition is the extraction of shape features. The extraction of object contours, the determination of their angles of orientation, and the extraction of segments and characteristic points can provide much useful information about the shape of the object. The shape factor can be selected as one of the features. The shape factor is defined as the ratio of the figure perimeter to the square root of its area. Other features that do not depend on the sizes and the orientation of the image can be used.

6.2.1 Features for Complex Shape Recognition

In the approach proposed below, shape recognition [15] is based on the analysis of the characteristic parts of the complex shape. Complex shapes, for example, oak and maple leaves, differ from each other in terms of the number and the nature of protrusions and cavities. In this case, the shape factor can differ insignificantly. Therefore, for the recognition of complex shapes, it is important to determine the parameters of these characteristic parts and to determine the object shape according to these parameters.

As the shape features, let us select the characteristic parts of the objects, protrusions and cavities, and term them primary features. Below, we describe the algorithm of primary feature extraction. It is necessary to note that the algorithm was developed taking into account the convenience of its realization on the neurocomputer. The algorithm works with the binary image, i.e., the figure is represented on the image by the set of 1s and on the background by the set of 0s. To extract primary features, we create two additional figures: the first figure circumscribes the initial figure, and the second figure is inscribed in the initial figure. The primary features are the difference in the areas of the circumscribing figure and the initial one (in Fig. 6.6, the initial figure is presented in black and the circumscribing figure is depicted by horizontal lines), and also the difference in the areas of the initial figure and the inscribed one (in the same figure, the inscribed figure is in gray).

The construction of the inscribed and circumscribing figures is achieved by scanning the image by a circle of a specific radius. The circle of a radius r is defined by the uniformly distributed points (r = 10 pixels, n = 28, where n is a quantity of points). Figure 6.7 gives an example of several positions of this circle on the image of a truck. If the number of 1s (points belonging to the initial figure) that fell into the circle is lower than the threshold P (it is experimentally established that P = 13), then the circle center is excluded from the figure (Fig. 6.7, case 1). Several such iterations lead to the truncation of all convex parts. The figure remains as if it is

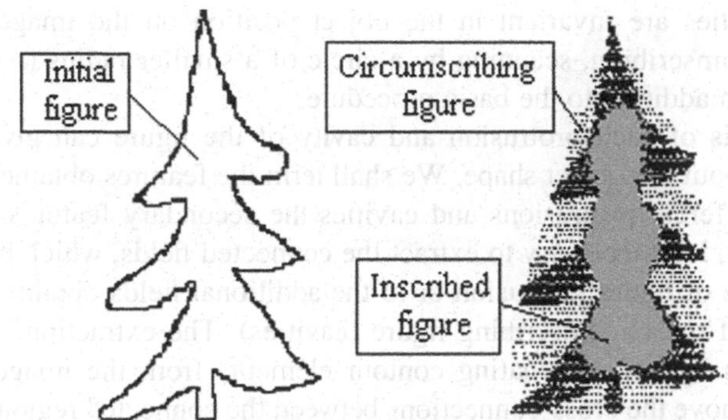

Fig. 6.6 Initial figure (black) and described and inscribed figures (horizontal lines and gray)

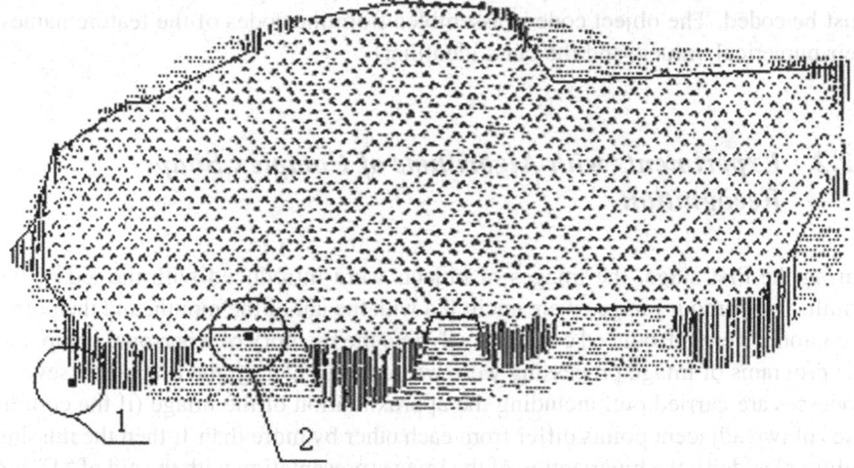

Fig. 6.7 The construction of the inscribed and circumscribing figures

inscribed into the initial image. Then the areas of initial and inscribed figures are calculated. The difference between them is used as the first feature for shape recognition.

Since the neurocomputer is oriented to work with the binary vectors of large dimensionality, the algorithm is built in such a way that, simultaneously, as many circles as pixels are counted in one image line (512 pixels in the last version of the neurocomputer). The extraction and the filling of cavities on the complex figure are done with the aid of the procedure of scanning by circle. The procedure must be repeated for the initial image. When a quantity of 1s in the circle exceeds a certain threshold ($P = 13$), the center of the circle is included in the figure (Fig. 6.7, case 2). Through several iterations, the initial image will be circumscribed by a certain figure whose points will fill its cavities. After calculating the areas of the initial and circumscribing figures and determining their difference, we will obtain the second feature for shape recognition. The processes of the truncation of protrusions and the filling of cavities are invariant in the object position on the image. To obtain smoother circumscribing, scanning by a circle of a smaller radius ($r = 8$, $n = 16$) is conducted in addition to the basic procedure.

The analysis of each protrusion and cavity of the figure can give additional information about the object shape. We shall term the features obtained during the analysis of different protrusions and cavities the secondary features. In order to calculate them, it is necessary to extract the connected fields, which belong to the truncated parts of figure protrusion or to the additional fields obtained during the construction of the circumscribing figure (cavities). The extraction of connected regions begins by our eliminating contour elements from the image, making it possible to remove the cross-connections between the connected regions to insulate them from each other. Subsequent analysis of each extracted area is carried out according to the same scheme as that for determining the primary features. When all connected regions are analyzed, the obtained primary and secondary features must be coded. The object code is assembled from the codes of the feature names, their numerical values, and the class identifier.

6.2.2 Experiments with Algorithms of Complex Shape Recognition

During the first stage, the images of objects were introduced with the aid of the mouth. The contour images were encircled with the aid of the mouth, and the points of contours were introduced consecutively into the memory of a personal computer. The programs of image processing were written in *C*. During the first stage, several processes are carried out, including the approximation of the image (if the coordinates of two adjacent points differ from each other by more than 1, then the missing points are added), the binarization of the image (presentation with the aid of "1" and "0"), and the filling of the figure's field with 1s.

Then, the values of primary features for the input image were calculated. From these values we formed two files: the first was for neural network training and the second for neural network examination. Feature coding for the first file was accomplished in the manner described in a previous chapter, i.e., the code vector contains the mask of the class identifier. The second file contains the code vector without the class identifier.

During the first stage, the neural network was trained for two classes of objects: the class of poplar trees and the class of fir trees drawn by hand. The features of ten representatives of each class of objects were input. The size of the associative field was 8,192 neurons, and the size of the ensemble was 256 neurons. For recognition, the objects were input without indication of their class. As a result, correct recognition was about 92%.

Another series of experiments was done with four classes: passenger automobiles, trucks, buses, and tanks. The images were drawn by hand. The quality of recognition was 90%. Thus, the possibility of complex shape recognition was shown with the aid of the neural network. During the imitation of the neural network on an IBM PC AT, recognition time was near 1 min. Using the neurocomputer NIK developed at the Insitute of Cybernetics of National Academy of Science, Ukraine, recognition time can be fractions of a second.

6.3 Recognition of Handwritten Symbols and Words

Before discussing the recognition of letters and words, let us introduce the term "small object," which we will use to define images of objects that occupy a small area (from several to several hundred pixels^2). A small object can be the component element of a texture (for example, the element of ornament) or the letter of text.

We see basic differences in the recognition of textures and small objects in that textures, as a rule, do not have a significantly high mutual arrangement of the elements of textures (tekstons), but have only their presence and quantity. During the recognition of small objects, the mutual arrangement of the elements plays a large role. Similarity in the recognition of textures and small objects, in our opinion, consists of the fact that recognition is accomplished "in the moment," i.e., it does not require the sequential procedures of image analysis.

We consider the recognition of separate symbols and words as the first stage of the recognition of handwritten texts. We assume that the solution to the problem of handwritten text recognition can be broken into two basic stages: 1) the recognition of the limited dictionary of handwritten words; 2) the recognition of arbitrary handwritten texts. The first task is considerably simpler because, if the word is considered a single object, it is possible to extract many more informative features than during the recognition of the separate letters that form the word. Therefore, during the recognition of words from the limited dictionary in the first stages, it is possible to achieve better results. This task can have independent value during the creation of devices for the readout of shapes, addresses on envelopes, and so on. For recognition of arbitrary words or word shapes, more complex algorithms with the use of knowledge bases and hierarchical recognizing structures will be required.

Let us discuss in detail the development of the system for word recognition from the limited dictionary. In an upcoming chapter, we discuss the hardware support of neural networks. We also examine the algorithms for the recognition of handwritten symbols. The neurocomputer was developed together with the Japanese company WACOM and named B-512. This work was executed by the authors together with D. A. Rachkovskij.

6.3.1 Features Utilized for the Recognition of Handwritten Words

We will denote the characteristic image elements, components of the letters, as features. Features can be segments of straight lines with different slope angles, arcs with different radii turned at a certain angle, the intersections of segments, and so forth. Each feature could be present or absent at the particular point of the image. For feature k, the value $y_k(ij) = 1$ is introduced if the feature is present at the point (i, j) of the image, and $y_k(ij) = 0$ is introduced if it is absent.

For the recognition of handwritten words, we used as features the segments of straight lines with specific lengths and orientations. In total, we used ten features. To extract these features, the lines of handwritten symbols are thinned to one pixel and then thickened to three pixels. Thus, the image of the symbol acquires the standard thickness of lines. The presence of a feature is determined by the presence of the straight line with the assigned length and orientation in the word symbols. The separately written word was read by a scanner with a resolution of 300–400 binary points per inch.

A brief description of the algorithm for handwritten word recognition for the limited dictionary consists of the following. The image obtained with the scanner is treated with the following operations:

1. thinning the lines on the image;
2. thickening the lines on the image;
3. extracting informative features;
4. coding the features;
5. inputting the obtained codes into the neural network for training or recognition.

To realize these operations, we used the algorithms of cellular logic, which were developed and tested by the authors [16].

Let there be an image window with the size of $3 * 3$ pixels. The pixel coordinates in the window at point (i, j) are shown in Fig. 6.8. Let us describe the algorithms of cellular logic.

6.3.1.1 The algorithm of line thinning

The algorithm is realized according to the following formulas:

$$\begin{aligned}
x_{ij} &= (x_{ij} \,\&\, (x_{ij-1} \,\mathrm{U}\!\sim\! x_{ij+1}))\,\mathrm{U}\,(x_{ij} \,\&\, x_{i-1j-1} \,\&\!\sim\! x_{i-1j})\,\mathrm{U}\,(x_{ij} \,\&\, x_{i+1j-1} \,\&\!\sim\! x_{i+1j}),\\
x_{ij} &= (x_{ij} \,\&\, (x_{ij+1} \,\mathrm{U}\!\sim\! x_{ij-1}))\,\mathrm{U}\,(x_{ij} \,\&\, x_{i-1j+1} \,\&\!\sim\! x_{i-1j})\,\mathrm{U}\,(x_{ij} \,\&\, x_{i+1j+1} \,\&\!\sim\! x_{i+1j}),\\
x_{ij} &= (x_{ij} \,\&\, (x_{i-1_j} \,\mathrm{U}\!\sim\! x_{i+1_j}))\,\mathrm{U}\,(x_{ij} \,\&\, x_{i-1j-1} \,\&\!\sim\! x_{ij-1})\,\mathrm{U}\,(x_{ij} \,\&\, x_{i-1j+1} \,\&\!\sim\! x_{ij+1}),\\
x_{ij} &= (x_{ij} \,\&\, (x_{i+1j} \,\mathrm{U}\!\sim\! x_{i-1_j}))\,\mathrm{U}\,(x_{ij} \,\&\, x_{i+1j-1} \,\&\!\sim\! x_{ij-1})\,\mathrm{U}\,(x_{ij} \,\&\, x_{i+1j+1} \,\&\!\sim\! x_{ij+1}),
\end{aligned} \tag{6.5}$$

$$\text{where } x_{ij} = \begin{cases} 1, & \text{if the pixel with coordinates } ij \text{ belongs to the line,} \\ 0, & \text{if the pixel with coordinates } ij \text{ does not belong to the line.} \end{cases}$$

x_{i-1j-1}	x_{i-1j}	x_{i-1j+1}
x_{ij-1}	x_{ij}	x_{ij+1}
x_{i+1j-1}	x_{i+1j}	x_{i+1j+1}

Fig. 6.8 Image window with the size of 3 * 3 pixels

The first formula from this group of formulas makes it possible to thin the line to the left, i.e., with the appearance of the situation $\begin{pmatrix} 011 \\ 011 \\ 011 \end{pmatrix}$, this formula's application will lead to the situation $\begin{pmatrix} 011 \\ 001 \\ 011 \end{pmatrix}$. The second formula thins the line to the right, and the third and fourth formulas work respectively from the top and bottom.

All operations in the described algorithm are done on the neurocomputer in parallel for the row consisting of 512 symbols. The entire image is processed via the sequential treatment of such rows. The operations enumerated in the formulas are performed consecutively. The algorithm is terminated in cases where, as a result of applying the operations on the entire image, a quantity of unit elements remains constant.

A simplified version of line thinning was developed and tested for the purpose of accelerating the algorithm's work. In this version, only three pixels were examined. Let us examine the algorithm of thinning from the left, right, top, and bottom. The algorithm of line thinning is composed from all four algorithms. Let us examine thinning from the left, i.e., when we have the following situation of pixels (0 1 1) and it is possible to remove the central unit:

$$x_{ij} = x_{ij} \;\&\sim (x_{ij+1} \;\&\; x_{ij} \;\&\sim x_{ij-1}). \tag{6.6}$$

In the situation where pixels are located vertically, $\begin{pmatrix} 0 \\ 1 \\ 1 \end{pmatrix}$, it is possible to carry out thinning from the top. The following formula is used for this purpose:

$$x_{ij} = x_{ij} \;\&\sim (\sim x_{i-1j} \;\&\; x_{ij} \;\&\; x_{i+1j}) \tag{6.7}$$

Similarly, for thinning from the right (situation (1 1 0)) and from below (situation $\begin{pmatrix} 1 \\ 1 \\ 0 \end{pmatrix}$), the formulas are as follows:

$$\begin{aligned} x_{ij} &= x_{ij} \;\&\sim \left(\sim x_{ij-1} \;\&\; x_{ij} \;\&\sim x_{ij+1}\right), \\ x_{ij} &= x_{ij} \;\&\sim \left(\sim x_{i+1j} \;\&\; x_{ij} \;\&\; x_{i-1j}\right). \end{aligned} \tag{6.8}$$

6.3.1.2 Algorithm of line thickening

Line thickening is carried out according to the formula:

$$x_{ij} = x_{ij} U x_{i-1j} U x_{i+1j} U x_{ij-1} U x_{ij+1}. \tag{6.9}$$

The thickening cycle is carried out either one or two times depending upon the result of an experimental test of the recognition algorithm. The operations of line thickening on the neurocomputer are performed simultaneously with the 512-bit rows.

Extraction of informative features

Each feature is determined by the set of the coordinates of unit elements that form it. For example, a diagonal straight line from five pixels can be done with the set of coordinates {(−2, −2) (−1, −1)(0, 0)(1, 1)(2, 2)}. The coordinate origin is selected at the center of the feature. For feature extraction, it is convenient to use parallel bitwise operations of the neurocomputer B-512. To extract the feature at the point having coordinates (r, s) in the image, we use the following formula:

$$y = x_{r-i_1,s-j_1} \,\&\, x_{r-i_2,s-j_2} \,\&\, \ldots \,\&\, x_{r-i_m,s-j_m}, \qquad (6.10)$$

where i_p, j_q are coordinates from the coordinate set of feature k; and x is the binary value of brightness. This formula reflects the calculations carried out for one point. During the use of bitwise operations of the neurocomputer, the calculation for the row of 512 pixels is produced simultaneously.

The selection of informative features can be executed with the aid of different algorithms. The simplest method is to select the features encountered most frequently in the training samples. More developed algorithms calculate the Shannon measure of the feature entropy. However, we assume that the most effective method of extracting informative features gives an evolutionary algorithm. For the execution of an evolutionary algorithm, a very high operation speed of the recognition device is necessary. With the neurocomputer B-512, we realize this algorithm. The algorithm of evolutionary optimization and its realization with the aid of the neurocomputer will be described below.

Information coding

The basic principle that distinguishes our neural network from other analogous networks is that any element that carries some information is coded in our network not by separate neurons but by certain subsets of neurons. Thus, during pattern recognition in our network, if any feature must be fed to the entrance of the network, this leads to the forced excitation of many neurons corresponding to this feature. In order to pass from the usual methods of information coding to coding with the aid of neuron subsets, we developed special coding procedures, which will be described below.

Coding of binary features

Let us describe the binary feature as any feature that could be present or absent in an object but for which it is not possible to indicate its numerical value. For example, for the binary feature "This word starts with the letter 'B'," the word "Blue" will equal 1 and the word "Green" will equal 0. Let a certain class of objects be described by the set of binary features $U = (u_1, \ldots, u_k)$, and let each object from this class be represented in the form of the binary vector in which the component i equals 1 if the corresponding feature is present in the object.

Let the neural network utilized for recognition have n neurons. The presence of the feature i is coded by the excitation of m neurons. For each feature, the neurons that are excited upon the appearance of this feature are selected with the aid of a random procedure. We introduce a binary vector that contains "1" corresponding to the excited neurons and "0" corresponding to the non-excited neurons. Let us term this binary vector the mask of the corresponding feature and denote the mask of feature i with M_i. If the object contains several features i, j, k, then the mask of Z is defined as a bitwise logical "OR" of the corresponding masks:

$$Z = M_i \text{ U } M_j \text{ U } M_k. \tag{6.11}$$

Coding the feature position on the image

Let us examine the handwritten symbol located on the image in the window having the size $r * s$. Let x be the horizontal coordinate of the pixel and y be the vertical coordinate ($x = 1, \ldots, r$; $y = 1, \ldots, s$). For each value of x and y are introduced the corresponding masks M_x and M_y, which are the "neural codes" of the corresponding coordinate. Then the code of the feature i, which has on the image the coordinates x_i and y_i, is equal to

$$Z_{X_i, y_i} = M_i \ \& \ M_{x_i} \ \& \ M_{y_i}, \tag{6.12}$$

where & is the bitwise logical "AND."

The code of the handwritten symbol will be determined by the expression:

$$Z_c = \underset{i}{\text{U}} Z_{x_i, y_i}, \tag{6.13}$$

where U is step-by-step "OR" and is carried out for all features extracted on the image of the object.

To form the codes for x and y coordinates, a procedure is used that makes it possible to obtain the correlated masks. This means that the masks for coordinates differing little from each other have a large quantity of common unit elements, but the masks of the coordinates with more differences have few common unit

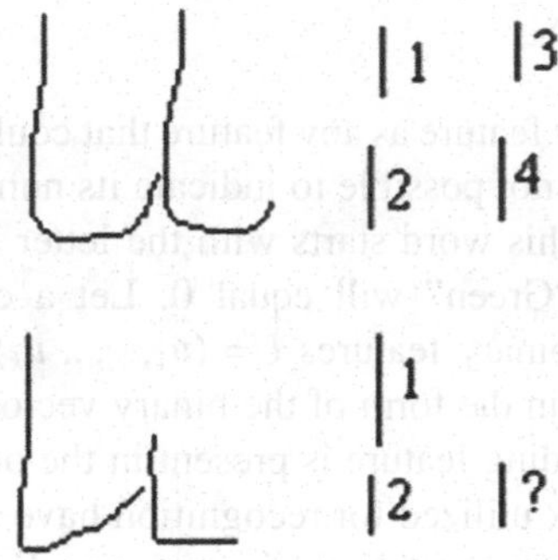

Fig. 6.9 Different ways of writing the letter "*u*"

elements. This coding takes into account the relative position of separate features. Using traditional methods of visual pattern recognition, there is a problem of repeated entry of the same feature in the description of an object. If the same feature is encountered several times, this requires the sorting out of algorithms, which slows down the recognition rate. Let us examine, for example, the handwritten letter *u* (Fig. 6.9).

Figure 6.9 depicts an example of the description of symbol "*u*" in the form of the features represented with line segments of different orientations (in Fig. 6.9, only vertical orientation is shown). Two versions of handwritten characters are given. Each line segment is characterized by the angle of the slope and position on the image. We will assume that the slope angle corresponds to the name of the feature (in the real tasks, we used ten different slope angles, which corresponded to ten different names). The position of each feature is given by two coordinates of the center of the corresponding segment.

Let us consider the task of recognition of the handwritten symbols described by this feature set. Let us assume that the method of potential functions is used for recognition. For this method, it is necessary to describe each concrete image by the fixed set of the parameters. In this case, difficulties are caused by the following. Firstly, some features can be absent in the image of a handwritten symbol; therefore, it is necessary to reach an agreement concerning what coordinates are assigned to the absent feature. Secondly, some features can be present in different quantities, so the problem here is how to order the set of features. It is very difficult to recognize and to fix different occurences of the same feature in the handwritten character description. Figure 6.9 shows an example of this due to the distortion of the handwritten symbol. In one case, the features 1, 2, 3, and 4 were extracted and in the other case, 1, 2, and ? were extracted. However, the system does not know the numeration of similar features, and in general, it is necessary to sort out all combinations of the correspondence of the feature numbers and to attempt to estimate the results of recognition for each combination. But since a quantity of features can reach several tens, it is practically impossible to sort out all combinations. In order to avoid this complex problem, it is expedient to use a coding method that would make it possible to recognize symbols independently of their position on the image. One appropriate method is shift coding.

Besides the input layer of neurons, the network also contains the output layer of neurons, which contains the masks of the names of the handwritten characters. Between the input and output layers are the connections whose weights are changed during training. Training is produced according to the results of the handwritten character recognition. If the neural network recognizes characters correctly, the weights do not change. But if the neural network makes errors, the weights of the connections leading from the excited neurons of the input layer to the neurons of the erroneously selected mask name decrease, and the weights of the connections leading to the neurons of the correct mask increase.

The experimental results

For these experiments created and conducted with D. A. Rachkovskij, ten handwritten words in English were selected, along with characters from the Latin alphabet and handwritten digits. During the recognition of handwritten words, the probability of correct recognition was 99–100% (for the familiar writer). During the recognition of Latin characters and handwritten digits, this probability was somewhat lower (98%). For the unfamiliar writer, the probability of handwritten word recognition was approximately 80%, and after the optimization of the set of features, it was about 84%. Let us consider in detail the optimization of the feature set.

Optimization of the feature set

The purpose of these experimental investigations is to improve recognition rate through optimization of the feature set. Ten features utilized for handwritten character recognition during the first stage of studies were selected intuitively. However, there is an approach that makes it possible to optimize the feature set. For this approach, an extended feature set containing many more features than the initial set is also formed intuitively. The problem is to select from the extended feature set the subset that optimizes a certain function of the quality of recognition. This problem was solved by the simulation of the evolution of biological species. The method of evolutionary simulation, or, as it is now called, the evolutionary algorithm, was proposed in the 1960s and at present is widely used by many authors [17–20].

Each "individual" is placed in correspondence with the feature subset that the "individual" is trained to recognize. Thereafter, the function of the quality of recognition is calculated. The process of creating the "offspring of the individual" is simulated using "mutations" and the process of "natural selection." We did not use crossing over in this work.

The feature set was optimized for the recognition of ten handwritten words written by different people. The extended feature set contained 41 features that are segments of lines and arcs of different length and radii having different orientations. Each of the features was encountered in the words, but a priori it is unknown what combinations of features make it possible to obtain better recognition results. The task of the evolutionary algorithm, which simulates biological evolution, is the search for such combinations.

Let us term "individual" the 41-bit binary vector $E = (e_1, e_2, \ldots, e_{41})$ in which each bit corresponds to one feature. If $e_i = 1$, the feature i is included in the subset for recognition, but if $e_i = 0$, the feature is not included in the subset. Let us define the mutation m_i as a change in the value of the component i of vector E (0 to 1 or 1 to 0). The "unisexual" evolutionary algorithm in which all "offsprings" are born to one "parent" but not to a pair of "parents" was used. Random mutations occurred. The probability of the mutation of each feature $p(m_i)$ depends only on the function of quality of the parent:

$$p(m_i) = p(Q), \tag{6.14}$$

where the quality function Q was defined as the fourth power of the error number of recognition,

$$Q = cN^4, \tag{6.15}$$

where N is the number of errors and c is a constant. The number of errors was calculated during recognition of different variants of ten handwritten words. In order to calculate quality function for a certain "individual," the neural network was trained to recognize words with the use of the feature subset of this "individual."

As the test, we selected the task of recognizing handwritten words written by different authors. The special shapes for writing English words (*one*, *two*, . . ., *ten*) were prepared. Each author prepared 12 versions of ten handwritten words. From this word set were formed two subsets, Q_1 and Q_2: Q_1 was for training (80 samples) and Q_2 was the control subset (40 samples). No training was done on the control samples. These samples were used for recognizing words and determining the percentage of correctly identified words. The words were input to the computer using a scanner with a resolution of 300–400 dots per inch.

For the simulation of natural selection, we used the algorithm in which the probability of the creation of "offspring" was proportional to the value of 1 / Q, allowing for an "individual" with the minimum quality function to generate more "offsprings" than an "individual" with a large quality function. The initial individuals were generated with the aid of a random-number generator.

The associative-projective ensemble neural network selected for the test of the evolutionary algorithm consisted of 4,096 neurons (experiments were also conducted with the network that consisted of 8,192 neurons, but only the first version will be described here). The coded input vector was supplied to the entrance of the network while the mask of the word was decoded in the output of the network. Connections between the neurons belonging to the input vector and the neurons forming the mask of the word were established in the matrix of the neuron connections. During recognition, the network's response was formed by the greatest intersection of the output vector with the masks of words. The neural network model was given by D. A. Rachkovskij for the test of the algorithm of evolutionary optimization.

The extraction of the subset of features S_1 as noted above corresponds to the construction of a certain recognition device that uses, for handwritten word recognition, only those features that enter into subset S_1. The initial set of such recognition devices was composed of the genotype of 0 generation. The probability of the unit elements in each bit of the "genotype" was equal to 1/2, i.e., in each vector of the genotype, approximately 20 features out of 41 were present (on average). We conducted three groups of experiments with different sizes of "generations." Each generation in our experiments included ten "individuals," 24 "individuals," or 32 "individuals." Subsequently, we will consider the experiments with the size of the generation equal to 32 "individuals" and ten "individuals."

Knowing the feature subset of an "individual," it is possible to form input vectors for each of the handwritten words and to use them to train the associative-projective neural network. After training the network for all words from training set Q_1, an examination is conducted on the test set Q_2, and the number of incorrectly identified words is calculated. Thereafter, the values of quality function Q are calculated. In our experiments, the quality function was:

$$Q = (c_1 * X + c_2 * Z)^{c_4}, \quad (6.16)$$

where c_1, c_2, c_4 are constants; X is a quantity of features; and Z is a quantity of errors during the samples of test set recognition.

Most of the experiments were carried out with c_1 equaling 0. In this case, the quality function was the function of a quantity of errors, i.e., the purpose of evolutionary optimization is the selection of "individuals" that decrease the number of errors. In the experiments, $c_4 = 4$. After calculating the quality functions Q for all "individuals" of 0 generation $G(0)$, the new generation $G(1)$ was formed. The formation of a "generation" occurs with the application of the stochastic procedure with which the "individuals" of the generation $G(0)$ generate "offsprings," i.e., the "individuals" of generation $G(1)$. The probability of creating new "individuals" is determined by the probability of er_l. The smaller the value of Q, the greater the probability that this "individual" generates "offsprings." The creation of "individuals" of a new generation is achieved in accordance with the formula $ver_1 = c * (1 / Q)$.

The probability of mutations is determined by the value of the probability of $vmut_1$. This probability is higher for "individuals" with a higher quality function value. The new generation also passed training and test procedures. This cycle of producing new generations is repeated until the definite value of the quality function is reached or the number of generation cycles defined previously is achieved.

In the first experiment, 11 generations were tested. The size of each generation was 32 "individuals." The average values of errors and of quality functions were calculated for each generation. Figure 6.10 gives the dependences of the average values of quality functions for each generation. The ten best "individuals" were extracted. Data about these ten "individuals" are presented in Table 6.1. The set of

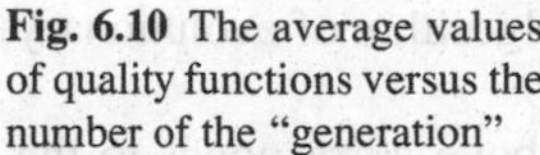

Fig. 6.10 The average values of quality functions versus the number of the "generation"

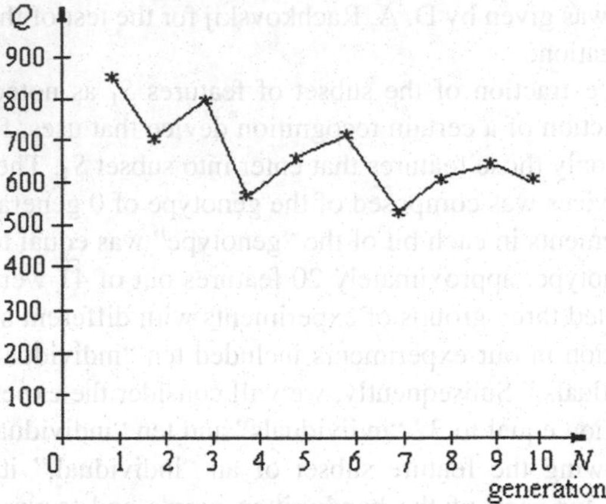

Table 6.1

N	N of generation	N of individuals	Values of quality functions	Error number
1	10	19	160	18
2	10	7	160	18
3	10	5	160	18
4	9	3	160	18
5	7	24	160	18
6	6	21	160	18
7	3	7	160	18
8	1	8	160	18
9	0	2	198	19
10	0	3	198	19

features corresponding to the best "individuals" was considered the best to use for subsequent work.

In the second experiment, we tested 11 generations, each having ten "individuals." Figure 6.11 represents the dependences of the average values of quality functions for each generation. The ten best "individuals" from all generations were selected. Their parameters are given in Table 6.2.

Results of recognition with these ten best "individuals" were tested on an associative-projective neural network that contains 8,192 neurons. The quantity of errors on the new network is given in Table 6.3.

A neurocomputer (to be described in detail in Chapter 7) with high productivity developed by Ukrainian and Japanese scientists from the company WACOM was used in the experiments. The complete cycle of training and recognition required several minutes. The results show that the evolutionary model makes it possible to improve the feature set, but neurocomputer productivity was insufficient for a detailed study of the properties of the evolutionary optimization algorithm. For this purpose, neurocomputer productivity must be increased in such a way that the cycle of training and recognition is approximately 1 second (or less). There are technical capabilities for the realization of such a neurocomputer [21, 22].

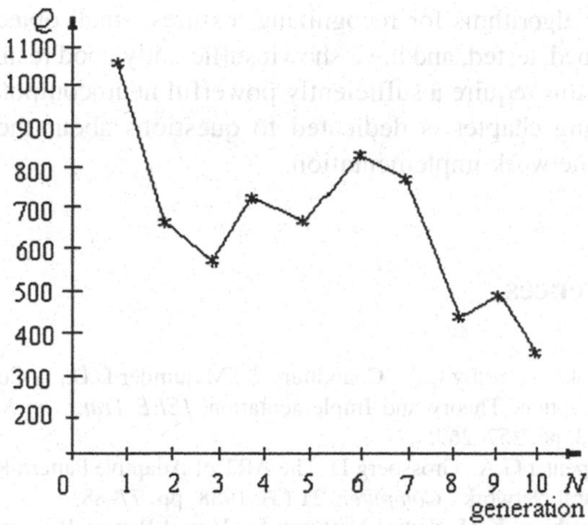

Fig. 6.11 The average values of quality functions versus the number of the "generation"

Table 6.2

N	N of generation	N of individuals	Values of quality functions	Error number
1	20	1	77	15
2	8	1	100	16
3	7	3	127	17
4	9	3	127	17
5	12	8	160	18
6	13	6	160	18
7	0	9	198	19
8	3	6	198	19
9	4	1	198	19
10	7	0	198	19

Table 6.3

N	1	2	3	4	5	6	7	8	9	10
Error number	23	17	18	21	21	23	23	21	22	16

6.4 Conclusion

Associative-projective neural networks present an effective means for texture and small image recognition. Neurocomputers developed for the realization of associative-projective neural networks proved to be highly effective not only for neural network implementation but also for the fulfillment of the preliminary processing of binary images.

The genetic algorithms that simulate the evolution of biological species are an effective means for selecting informative features for image recognition. Highly productive neurocomputers are required for the realization of genetic algorithms.

The algorithms for recognizing textures, small objects, and object shapes were developed, tested, and have shown sufficiently good results. However, the developed algorithms require a sufficiently powerful neurocomputer for their realization. The following chapter is dedicated to questions about the creation of hardware for neural network implementation.

References

1. Basak J., Murthy C.A., Chaudhury S., Majumder D.D., A Connectionist Model for Category Perception. Theory and Implementation, *IEEE Trans. on Neural Networks*, Vol. 4, No. 2, 1993, pp. 257–269.
2. Carpenter G.A., Grossberg D. The ART of Adaptive Pattern Recognition by a Self-organizing Neural Network , *Computer*, 21 (3), 1988, pp. 77–88.
3. Fukushima K. A Neural Network for Visual Pattern Recognition, *Computer*, 21 (3), March 1988, pp. 65–75.
4. Kohonen T., Speech Recognition Based on Topology-Preserving Neural Maps. Neural Computing Architectures. The Design of Brain-Like Machines. Ed. I. Aleksander. North Oxford Academic Publishers, London, 1989, pp. 26–40.
5. Lippmann R.P. A Critical Overview of Neural Network Pattern Classifiers, Proceedings of IEEE Conference, Paris, France, 1991, pp. 266–275.
6. Marr D. Vision. Freeman, San Francisco, 1982.
7. Julesz B. Experiments in the Visual Perception of Texture, *Sci. Am.*, 1975, 232, pp. 34–27.
8. R. Schwartz, Patent 4433912, USA, G 05 B 19/405. Method and a Circuit for Determining a Contour in an Image, Publ. 28.02.1984.
9. Amosov N.M., Baidyk T.N., Goltsev A.D. et al., Neurocomputers and Intelligent Robots, Ed. Amosov N.M., Kiev: Nauk. Dumka, 1991, p. 272 (in Russian).
10. Duda R., Hart P. Pattern Classification and Scene Analysis. New York: Wiley-Interscience Publ., 1973, pp. 482.
11. Marr D., Poggio T. Cooperative Computation of Stereo Disparity, *Science*. 194, 1976, pp. 283–287.
12. Aizerman M.A., Braverman E.M., and Rozonoer L.I., Theoretical Foundations of the Potential Function Method in Pattern Recognition Learning. *Automation and Remote Control*, 25, 1964, pp. 821–837.
13. Davies E.R. Application of the Generalised Hough Transform to Corner Detection, IEEE Proc., E135, No. 1, 1988, pp. 49–54.
14. Leavers V.F., Boyce J.F. An Implementation of the Hough Transform Using a Linear Array Processor in Conjunction with a PDP/11, Nat. Phys. Lab. Div. Inf. Technol.and Comput. Rept., No. 74, 1986, pp. 1–11.
15. Kussul' E.M., Baydyk T.N. Design of a Neural-Like Network Architecture for Recognition of Object Shapes in Images. *Soviet Journal of Automation and Information Sciences (formerly Soviet Automatic Control)* (ISSN 0882-570X), Vol. 23, No. 5, 1990, pp. 53–59.
16. Baidyk T.N., Kussul' E.M., 1999, Application of Genetic Algorithms to Optimization of Neuronet Recognition Devices. *Cybernetics and System Analysis*. Vol. 35, No. 5, pp. 700–707.
17. Clopf A.H., Gose E.E. An Evolutionary Pattern Recognition Network, *IEEE Trans. Syst. Sci. and Cybernet*, Vol. 5, No. 3, 1969, pp. 247–250.
18. Ferri F.J., Pudil P., Hatef M., Kittler J. Comparative Study of Techniques for Large-scale Feature Selection, Pattern Recognition in Practice IV, Vlieland, June 1–3, 1994, pp. 43.
19. Kussul E.M., Luk A.N. Evolution as a Process of Search for an Optimum, Soviet Science Review. Scientific Developments in the USSR. IPC Business Press Ltd., Vol. 3, No. 3, 1972, pp. 168–172.

20. Mucciardy A.N., Gose E.E. Evolutionary Pattern Recognition in Incomplete Nonlinear Multithreshold Networks, *IEEE Trans. Electron. Computers*, EC–15, No. 2, 1966, pp. 257–261.
21. Kussul E.M., Rachkovskij D.A., Baidyk T.N. Associative-Projective Neural Networks: Architecture, Implementation, Applications, Proceedings of Fourth Intern. Conf. "Neural Networks & their Applications", Nimes, France, 1991, pp. 463–476.
22. Kussul E.M., Baidyk T.N., Rachkovsky D.A. Neural Network for Recognition of Small Images, Proceedings of the First all-Ukrainian Conference UkrOBRAZ'92, Kyjiv, Ukraine, 1992, pp. 151–153.

Chapter 7
Hardware for Neural Networks

The simulation of neural networks on conventional computers is time-consuming because parallel structures and processes are transferred to a sequential machine. New computational means oriented to the realization of neural network paradigms were developed, these are called neurocomputers. Universal and specialized neurocomputers [1–7] were created. In this chapter, several versions of hardware for the proposed neural networks will be considered.

7.1 Neurocomputer NIC

The purpose of creating the neurocomputer was to increase the speed and expansion of simulated neural networks. At the Institute of Cybernetics of the National Academy of Sciences, Ukraine, under the management of the Doctor of Sciences E. M. Kussul, the first neurocomputer for associative-projective neural networks was developed and tested. The block diagram of the neurocomputer, naed NIC, is given in Fig. 7.1.

The model consists of two basic blocks: the block of processing modules and the control unit. The former contains the random access memory (RAM), arithmetic-logical unit (ALU), multiplexer (MUX_1), shift register (SRG_1) and unit of binary counters (St_1). The latter contains the high-speed memory unit (CM), arithmetic-logical unit (ALU_2), shift registers (SRG_2 and SRG_3), multiplexer (MUX_2), binary counter (Ct_2), training address register (IRG), and synchronizing circuit (not shown in the figure).

7.1.1 Description of the Block Diagram of the Neurocomputer

The control unit is built as a computer with RISC architecture. It must form instructions for the processing block, ensuring the imitation of the work of the

DOI 10.1007/978-3-642-02535-8_7, © Springer-Verlag Berlin Heidelberg 2010

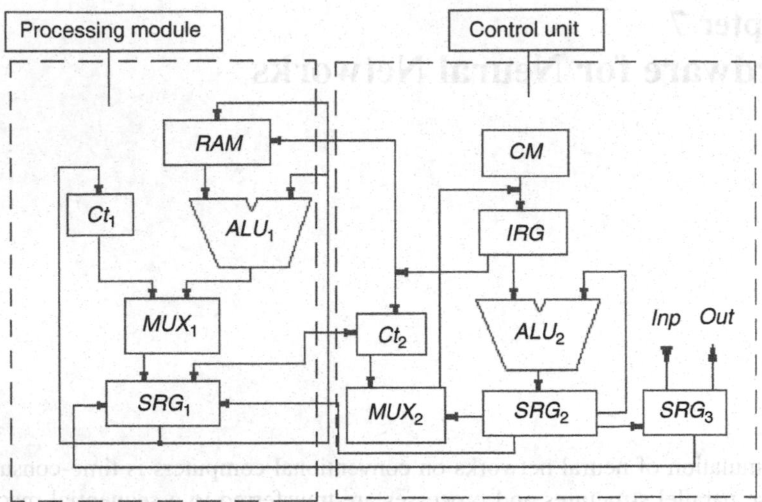

Fig. 7.1 The block diagram of neurocomputer NIC

neural network. Furthermore, it serves as a connection with the personal computer. The processing block is the multi-bit logical processor augmented by the set of binary counters.

All operations connected with the simulation of buffer fields, the coders and decoders, and the training of associative fields are performed by a logical processor. Since all these operations can be done simultaneously for all neurons of one field, it is expedient to select the word length of the logical processor equal to the length of the neuron field. However, in practice this leads to excessive equipment expenditure. Therefore, when developing the neurocomputer, 256 bits were selected as the word length of the logical processor. The operations, which are connected with the calculation of the input excitations of neurons in the associative fields and the fulfillment of the threshold procedure, are accomplished by a set of binary counters (Ct_1). The binary counter Ct_2, connected with shift register SRG_1, serves for the rapid definition of the number of active neurons that is necessary while training associative fields and counting their activity.

The matrices of synaptic weights for the associative fields are stored in the random access memory (RAM). During training, the contents of these matrices change. At the end of the work session, the contents of matrices are rewritten on a hard drive. At the beginning of the new cycle, the contents of the matrices are loaded in RAM. These operations are accomplished through shift registers SRG_2 and SRG_3. RAM is used not only for storing the synaptic weights of neuron connections, but also for storing the temporary values of excitation at inputs and outputs of the neurons and for storing the intermediate parameters obtained during the calculations. RAM contains a structure of the word with identical word length (for example, 256 bits). An arithmetic-logical unit (ALU) is used to fulfill the bitwise logical operations above the words of overall length (for example, 256).

The multiplexer of channels (MUX_1) is used to connect the outputs of the unit of binary counters (Ct_1) or the outputs of the arithmetic-logical unit (ALU_1) to the inputs of the shift register (SRG_1). The multiplexer makes it possible to write down into the shift register the results of the work of either binary counters or the arithmetic-logical unit.

Neurocomputer programming is accomplished in a specially developed language similar to Assembler. Software was developed that includes basic building blocks for the simulation of neuron networks. The emulator of this version of the neurocomputer was developed for checking the correctness of the neurocomputer's operation system, and it served for writing and testing neurocomputer programs in a medium convenient for programmers.

7.1.2 The Realization of the Algorithm of the Neural Network on the Neurocomputer

The neurocomputer works in different modes: the mode of the load, of the coding of input information, of training, of associative recall of information, and of decoding.

In the mode of the load, the simulation program of the neural network and the initial data necessary for the simulation are loaded from the peripheral device (for example, the PC), with bidirectional bus.

In the mode of information coding, parameters supplied to the neural network are transformed into a binary vector of size n, in which the unit elements are associated with the excited neurons of the network. The following conditions for coding are observed: a quantity of unit elements (m) in the vector code must have an approximately constant value that is many times less than the quantity of neurons in the network ($m << n$); the input parameters are coded by the subsets of the randomly selected unit elements, and the different statistically independent subsets (a small intersection of subsets is allowed) correspond to different parameters; and during the input of several parameters simultaneously, the resulting code is defined as the bitwise disjunction of the initial codes, which subsequently undergoes normalizing (removal of excess unit elements). Coding is done as follows: for each of the input parameters, the n-dimensional vector containing m unit elements is formed. This vector is recorded in RAM and is stored there during the entire work session of the neural network. Pseudorandom number generation is realized by a program using the array of the initial random numbers that are stored in RAM, the logical unit, the multiplexer of channels, and the shift unit.

When the set of parameters $x_1, \ldots, x_k$ (where k is a quantity of parameters) is presented to the neural network, the codes of these parameters $X_1, \ldots, X_k$ are extracted from RAM and are united disjunctively in the shift register. Obtained as a result, code $Y = (X_1 \cup X_2 \cup \ldots \cup X_k)$ is written to RAM. The code stored in the shift register is shifted cyclically to the randomly selected quantity of bits, after which the operation $Z = Z \,\&\, {\sim}Y$ is performed, where Z is the code stored in the shift register, and Y is the code stored in RAM; & is conjunction; and ~ is negation. This operation continues until, in code Z, a quantity of unit elements becomes less than m. The resulting vector is called the normalized input vector of the neural network. The process of input information coding is terminated.

If the neural network was not trained previously, then the training mode begins when the synaptic weights of interneuronal connections are reset to 0. After this, the sets of parameters that must be associative, and connected with each other, are supplied to the input of the neural network. Each of the input sets of the parameters is coded, and the code obtained in the shift register is maintained in RAM. Next, the process of training occurs, though we will not discuss this in detail. Note that the numbers of the synaptic matrix rows corresponding to active neurons are extracted according to the numbers of active neurons, and these numbers are converted into the addresses of RAM. Each obtained matrix row M_i undergoes the following operation:

$$M_i^* = M_i \mathrm{U} \ (Z \ \& \ P), \tag{7.1}$$

where P is the vector formed with the aid of the pseudorandom number generator. The obtained vector M_i^* is written in RAM with the same address where the vector M_i was stored. The experiments with this algorithm of training showed that, in the neural networks, structures are formed that correspond to the neural ensembles.

In the mode of associative recall of information, sets of input parameters in which one part coincides with one of the ensembles and another part is selected randomly as a noise component are supplied to the input. The task of the neural network is to restore the missing parameters from the noise set and to eliminate the noise component. Coding of the input set of parameters in this mode is accomplished exactly as in the mode of training. Associative restoration is achieved in the process of cyclic recalculation of neuron activity.

In the mode of decoding, the neural network must decode to what class the excited neuron ensemble corresponds, which is achieved as follows. The obtained binary vector of neuron activity is compared with all vector codes of the classes stored in RAM. If a quantity of coinciding elements exceeds a certain threshold, it is assumed that the code of this class is present in the resulting output vector of the neural network.

This neurocomputer was used for texture recognition as described earlier.

7.2 Neurocomputer B-512

As a result of work with the neurocomputer described above, the need arose to create a more productive neurocomputer with greater memory and an improved system of instructions. To explore the possibilities of the new version of the neurocomputer, the authors of this book developed a neurocomputer emulator, and later, the neurocomputer was developed with the participation of D. A.

Rachkovskij, M. E. Kussul, and the Japanese company WACOM. The neurocomputer was named B-512. Sadao Yamamoto, Masao Kumagishi, and Yuji Katsurakhira participated on the Japanese side of the neurocomputer's development.

7.2.1 The Designation of the Neurocomputer Emulator

Usually, computer emulators are used during the development of devices for checking the functionality of the developed device and for the development of software before the production of the device itself. During the creation of B-512's software arose difficulties connected with the fact that access to memory was complex, and the development of programs was carried out in a low-level language. Therefore, for program debugging, it was desirable to have open access to the memory. To eliminate this disadvantage, the neurocomputer emulator was developed, which was capable of debugging programs. Let us consider in detail the structure of the emulator.

7.2.2 The Structure of the Emulator

The emulation of the neurocomputer was conducted at the level of neurocomputer instructions and did not affect the level of separate signals. A change in the state of all registers and the state of the neurocomputer's memory were considered during emulation. The following types of instructions were used in the neurocomputer: 1) shifts; 2) flags set; 3) reading from and writing into memory with the fulfillment of logical operations; 4) the calculation of "1s"; and so on. Table 7.1 gives the list of B-512's instructions. The following abbreviations are used: DRAM[] is dynamic RAM; $SRAM_1[]$ is static memory for instructions; $SRAM_2[]$ is static memory for data; LR_2, LR_3 are registers of the logical modules; RG_2, RG_3 are registers of the control unit (CU); CNT_1, CNT_2, CNT_3, CNT_4 are counters. In Table 7.1, the following additional designations are accepted:

label is the marker placed in $SRAM_1[L]$;
var is the variable stored in $SRAM_2[M]$;
L is the low-order bits of the register RG_1 of the control unit (CU);
M is the average bits of the register RG_1 of the control unit (CU);
const is the constant placed in *L*.

The following bitwise operations are realized on the neurocomputer:

~A –*NOT A*;
A & B –*A AND B*;
A / B –*A OR B*;
A ^ B –*A XOR B*.

Table 7.1

Instruction code, (octal system)	Instruction mnemonics	Instruction
001	R_2	$LR_2 = DRAM[L+var]$
002	R_3	$LR_3 = DRAM[L+var]$
003	N_2	$LR_2 = \sim DRAM[L+var]$
004	N_3	$LR_3 = \sim DRAM[L+var]$
005	A_2	$LR_2 = DRAM[L+var]\ \&\ LR_3$
006	A_3	$LR_3\ \&= DRAM[L+var]$
007	O_2	$LR_2 = DRAM[L+var]\ LR_3$
010	O_3	$LR_3\ LR_3 = DRAM[L+var]$
011	X_2	$LR_2 = DRAM[L+var]\ {}^\wedge LR_3$
012	X_3	$LR_3\ {}^\wedge= DRAM[L+var]$
013	W_2	$DRAM[L+var] = LR_2$
014	W_3	$DRAM[L+var] = LR_2 = LR_3$
015	SL	$LR_3(+RG_2) <<= L\ (mod\ 512)$
016	SR	$LR_3(+RG_2) >>= L\ (mod\ 512)$
017	RS	$LR_3(+RG_2) >>= CNT_4$
020	CT	*Counter simulation*
021	NA	$DRAM[RG_3] = \sim DRAM[RG_3]$
040	FLG	*Set flag*
041	JMP	*Goto label*
023	JCO	*If() goto label*
043	JSR	$var = ++CNT_2;$ *goto label*
044	RET	*return to var*
045	NS_3	$RG_3 = \sim RG_3 + var$
050	S_3S	$var = RG_3$
051	SCS	$var = CNT_1$
052	L_{13}	$RG_3 = const$
053	LS_3	$RG_3 = var + const$
054	A_{13}	$RG_3 += const$
055	AS_3	$RG_3 += var$
056	A_1S	$var += const$
057	ASS	$SRAM_2[CNT_1] += var + const$
060	LC_3	$CNT_3 = var$
061	S_1S	$var = const$
062	SSS	$var = SMO$
063	LCS	$CNT_1 = var$
064	L_2S	$RG_2 = var$
065	LCS	CNT_1++
066	LC_4	$CNT_4 = var$

In the emulator, the results of the instructions (the state of registers, counters, and so on) were given on the basis of Table 7.2.

In Table 7.2, the following designations are accepted:

– there are no changes;

? the content is changed;

D input data;

0 set bit into 0;

1 set bit into 1.

Table 7.2

Code	RG_2	RG_3	CNT_2	CNT_3	CNT_4	FRG	SMO	OFL	LR_2	LR_3
040	–	–	–	–	–	1	–	–	–	–
041	–	–	–	–	–	–	–	–	–	–
023	–	–	–	–	–	–	–	–	–	–
043	–	–	–	–	–	–	–	–	–	–
044	–	–	–	–	–	–	–	–	–	–
045	–	D	–	–	–	–	–	?	–	–
050	–	–	–	–	–	–	–	–	–	–
051	–	–	–	–	–	–	–	–	–	–
052	–	D	–	–	0	–	–	0	–	–
053	–	D	–	–	–	–	–	?	–	–
054	–	D	–	–	?	–	–	?	–	–
055	–	D	–	–	–	–	–	?	–	–
056	–	?	–	–	–	–	–	?	–	–
057	–	?	–	–	–	–	–	?	–	–
060	–	–	–	D	–	–	–	–	–	–
061	–	–	–	–	–	–	–	–	–	–
062	–	–	–	–	–	–	–	–	–	–
063	–	?	D	–	–	–	–	–	–	–
064	D	–	–	–	–	–	–	–	–	–
065	–	–	?	–	–	–	–	–	–	–
066	–	–	–	–	D	–	–	–	–	–
001	–	?	–	–	–	–	?	?	D	–
002	–	?	–	–	–	–	?	?	–	D
003	–	?	–	–	–	–	?	?	D	–
004	–	?	–	–	–	–	?	?	–	D
005	–	?	–	–	–	–	?	?	D	–
006	–	?	–	–	–	–	?	?	–	D
007	–	?	–	–	–	–	?	?	D	–
010	–	?	–	–	–	–	?	?	–	D
011	–	?	–	–	–	–	?	?	D	–
012	–	?	–	–	–	–	?	?	–	D
013	–	?	–	–	–	–	?	?	–	–
014	–	?	–	–	–	–	?	?	D	–
015	?	–	–	–	0	–	?	–	–	?
016	?	–	–	–	0	–	?	–	–	?
017	?	–	?	–	0	–	–	–	–	?
020	–	?	–	0	?	–	?	?	?	?
021	–	?	–	0	?	–	?	?	?	?

FRG is the register of the flag used to interface with an IBM PC AT personal computer; SMO is a quantity of 1s in the register LR_2; and OFL is overflow in register RG_3. The special feature of B-512 is the use of the long registers, some of which contain 512 bits, for example, LR_1, LR_2, and LR_3. It was inconvenient to use digital representation to visualize the state of this register. Therefore, we proposed a graphic presentation of the state of the register on the personal computer's screen, in which 0 corresponds to the short vertical line and 1 to the long vertical line, and different bytes were reflected by different colors. With this image, it was possible to present the state of the register on the screen as one line. Based on the emulator, the debugger was created by Masao Kumagishi, a colleague at WACOM, and was used to prepare the programs for the preliminary processing of images and to extract features in the recognition programs for handwritten words and characters. We used the emulator to analyze different structures of neural networks.

7.2.3 The Block Diagram of B-512

The block diagram is represented in Fig. 7.2. CU is the control unit; $SRAM_1$ is the random access memory for storing the neurocomputer's programs; $SRAM_2$ is the random access memory for storing data; LMs are the logical modules; DRAM is the dynamic memory whose size for each module is 256 K x 512 bits; and PC is the personal computer, which enables interfacing with the neurocomputer.

The basic difference between the neurocomputer B-512 and the neurocomputer NIC is that the quantity of operations needed for the simulation of specific neural network functions is increased in B-512. The number of such operations includes the hardware procedure for the calculation of a quantity of unit bits in the vector of the activity of neurons.

The calculation of a quantity of 1s in the 512-bit vector is carried out with one instruction whose duration is approximately 1 μs, whereas with NIC, it is necessary to carry out 512 sequential shifts of the vector for this purpose, demanding about 100 μs. Another operation of this type is automatically searching for the numbers of active neurons and writing these numbers into the memory $SRAM_2$. These numbers are used to select the lines from the matrix of synaptic weights. The selection of the necessary line also is achieved with one instruction. B-512 uses a more developed system for addressing dynamic memory, which makes it possible to form the index registers that are necessary to organize the cycles.

One of the distinctive features of B-512 is the formation of virtual counters for determining the input activity of neurons. NIC makes it possible to use counters with depths not exceeding eight bits; the depth of the counters of B-512 is

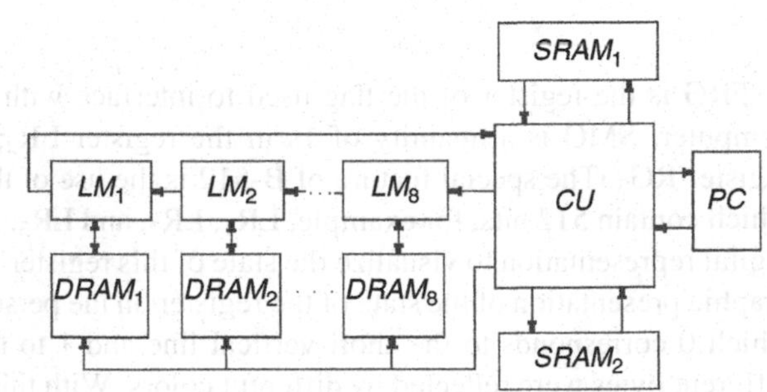

Fig. 7.2 The structure of the neurocomputer B-512

Fig. 7.3 The neurocomputer B-512

determined by the programmer and can reach 16 bits. Such counters permit work with neuron ensembles whose sizes can reach up to 32,000 neurons. To form this ensemble, the neuron field must contain hundreds of thousands of neurons.

The RAM of B-512 exceeds the memory of NIC by eight times and has 64 MB, which makes it possible to realize neural networks having up to 400 million synaptic connections. With the modular structure of the network and the sizes of each module of 1,000 neurons, it is possible to work with networks having up to 400,000 neurons. The neurocomputer's speed was 1 billion CUPS (connection updates per second). Let us note that the neurocomputer consumes only 3 W and includes 9 PCBs with the dimension of 145 x 170. D. A. Rachkovskij wrote the programs for base software for B-512. The following tasks were completed on the neurocomputer. Recognition of the separate symbols written by hand was executed by D. A. Rachkovskij. Recognition of handwritten words from the limited dictionary with the application of a genetic algorithm was executed by the authors. Figure 7.3. presents the photograph of B-512.

As a result of this work with B-512, the following requirements for the new devices of this class were formulated. First, it is necessary to increase the neurocomputer's memory. Second, the neurocomputer's speed must also be increased. This model of the neurocomputer cost $4,500 (price in 1992). Therefore, the third requirement is a reduction of the neurocomputer's cost. The fourth requirement is the creation of program interface in order to facilitate neurocomputer programming. All requirements are the usual ones addressed by developers who work with this type of hardware. However, the accomplished work generated a specific requirement. The fifth requirement consists of the need to develop architecture convenient for the realization of not fully connected neural networks.

7.3 Conclusion

The paradigm of associative-projective neural networks makes it possible to create simple and highly productive hardware for the simulation of neural networks (neurocomputers). Modern electronics make it possible to create neurocomputers for the simulation of neural networks containing hundreds of thousands of neurons and hundreds of millions of interneuronal connections. One way to increase neurocomputer speed is to create hardware for the not fully connected structure of associative-projective neural networks.

References

1. Kussul E. M., Lukovitch V. V., Lutsenko V. N., Multiprocessor Computers for Control of Transport Robots in Natural Environment, *USiM*, No. 5, 1988, pp. 102–104.
2. Drumheller M., Poggio T., On Parallel Stereo, Proceedings of IEEE Conference Rob. and Autom., San Francisco, Calif., Apr. 7–10, 1986, 3, pp. 1439–1448.
3. Hopfield J.J. Neural Networks and Physical Systems with Emergent Collective Computational Abilities, Proc. Net. Acad. Sci. USA, 1982, 79, pp. 2554–2558.
4. Webster W.P. Artificial Neural Networks and their Application to Weapons, *Naval Engineers Journal*, 1991, pp. 46–59.
5. Kussul E. M., Rachkovskij D. A., Baidyk T. N., 1991, Associative-Projective Neural Networks: Architecture, Implementation, Applications. In: Neural Networks & Their Applications, Nimes, France, EC2 Publishing, pp. 463-476
6. Baidyk T. N., Neural Network Subsets to Resolve Some Problems of Artificial Intelligence, Thesis for a Doctor's degree (D. Sc. Eng), Institute of Cybernetics of Ukraine, 28.09.1994.
7. Baidyk T. N., Neural Networks and Artificial Intelligence Problems. Kiev, Naukova Dumka, 2001, p. 264 (in Russian).

Chapter 8
Micromechanics

We suggest a new technology for the production of low-cost micromechanical devices that is based on the application of microequipment, similar to conventional mechanical equipment, but much smaller. It permits us to use the conventional technology for the mechanical treatment of materials and for the automatic assembly of mechanical and electronic devices for manufacturing micromechanical and microelectromechanical devices of submillimeter sizes. We call it “Microequipment Technology” (MET).

8.1 The Main Problems of Microfactory Creation

MET will use microequipment for manufacturing commercial products and in turn will produce the necessary microequipment units. The decrease in manufacturing costs of microdevices will be achieved on the basis of mass parallel production processes used in MET [1], instead of the batch processes used in Microelectromechanical Systems (MEMS) [2–4]. In accordance with MET, sequential generations of microequipment are to be created (Fig. 8.1) [5, 6].

The micromachine tools of the first generation could have an overall size of 100–200 mm. These sizes correspond to those of the Japanese microfactory machine tools [7]. These machine tools could be produced by conventional mechanical equipment. Using the first generation microequipment, it is possible to produce micromachine tools, microassembly devices, and so on of the second generation. After that, each new generation of microequipment could be produced using the microequipment of the previous generations and should have microequipment smaller than that of previous generations. The microequipment of each generation could be produced from low-cost components. The required precision could be obtained due to a decrease of the machine tool size and a use of adaptive control algorithms. So, low cost and the possibility of scaling down components are the main demands of the proposed method of microequipment creation. On the basis of this method, we began to develop prototypes of the microequipment. In 1997, the

E. Kussul et al., *Neural Networks and Micromechanics*,
DOI 10.1007/978-3-642-02535-8_8, © Springer-Verlag Berlin Heidelberg 2010

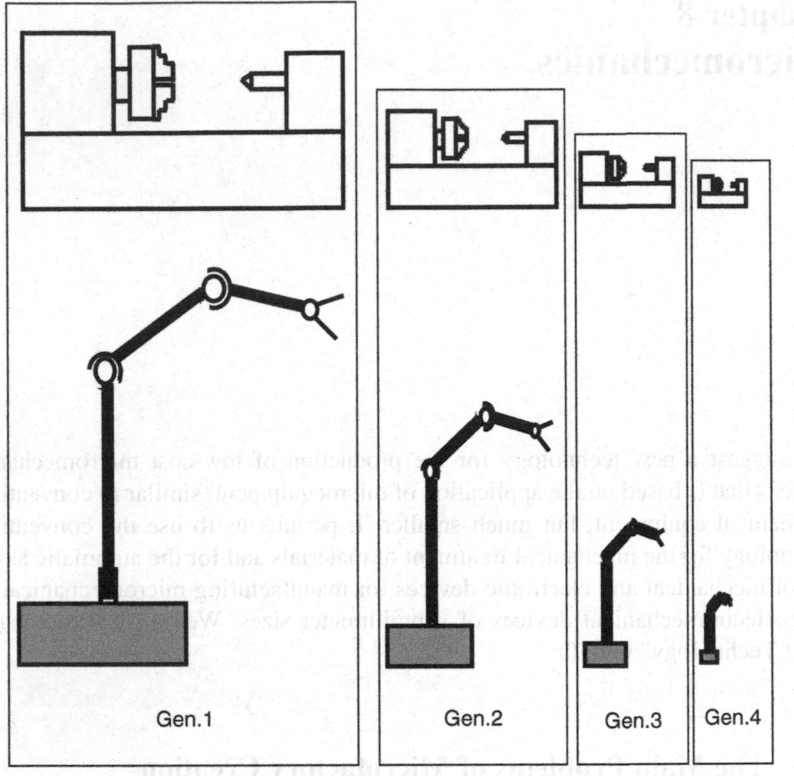

Fig. 8.1 The sequential generations of microequipment

first prototype of the first-generation micromachine tool was made and tested. The second prototype of the first-generation micromachine tool was made in 2000 [8]. The first generation contains micromachine tools and micromanipulators having an overall size of 100–200 mm. The second generation will contain micromachine tools and micromanipulators having an overall size of 50 – 100 mm and can be made with the first-generation microequipment. Each following generation would have smaller devices than the previous generation and could be made using the previous generation.

Micromechanical and MEMS devices are widely used for different applications. The automotive and aerospace industries use MEMS for producing sensors and actuators [9–17]. In adaptive optics and telecommunications, MEMS are used for optical switches [18–20, 17]. A very interesting application of MEMS is the production of displays based on arrays of micromirrors reflecting external light beams on to corresponding pixels of the screen [21]. On the basis of MEMS, new components for radio-frequency (rf) electronics [22–24] have been developed. MEMS permits improvements in the characteristics of heat exchangers [25–28] and is used widely in biotechnology [17, 29], medicine

[3, 17, 30, 31], neurophysiology [3], and so on. There are projects with the aim of developing mobile microrobots [32–34] and producing them using micromechanics technologies. Peripheral Research Corp., Santa Barbara, Calif. estimated the micromechanical devices market to grow to $11 billion in 2005 from between $3.7 billion and $3.8 billion in 2001, even as average sensor prices declined sharply in 2005 to $1 from $10 in 2001 [35].

At present, the base technology for micromechanical device production is lithography [3, 36]. Up to 99% of commercial MEMS production uses this technology [3], but lithography makes it possible to produce only two-dimensional (2D) shape details. The production of three-dimensional (3D) shape details has many problems, which is why many researchers try to use conventional methods for the mechanical treatment of materials to produce microdevices [3, 36–40]. For this purpose, they use precise and ultra-precise machine tools because tolerances of microdetails must be very small. The preciseness of machine tools may be determined as a relation of the machine tool's size to the minimal tolerances it ensures. To increase the preciseness of machine tools for microdetail manufacture, it is necessary to miniaturize them. Two main reasons for miniaturization of machine tools have been given [36]. The first is a decrease of heat deformation of machine tools with a decrease in their sizes. The second is a decrease of material consumption for machine tool production; in this case, more expensive materials with better properties can be used for machine tool manufacture.

There are also other reasons for miniaturization of machine tools, one being that the vibration amplitudes of small machine tools are lower than those of large ones because the inertial forces decrease as the fourth power of the scaling factor, and the elastic forces decrease as the second power of the scaling factor [41]. Moreover, smaller machine tools demand less space and lower energy consumption.

Microequipment technology is also advantageous because it preserves conventional mechanical technology methods in small-scale device manufacturing, which can lead to decreasing not only the cost of the products but also the time it takes for products to reach the market. In contrast, almost all MEMS-produced microdevices demand new mechanical designs. To invent and to develop a new device design requires a great deal of time. Special kinds of materials, which are used in MEMS technology, demand additional investigations of the microdevice properties and the resolution of many problems (such as wearing, lubrication, etc.), which are solved in macromechanics for traditional materials. In comparison with MEMS, microequipment technology permits us to make minimal changes in the design and production methods of microdevices.

A special project for microfactory creation based on miniature micromachine tools has been initiated in Japan [42]. The Mechanical Engineering Laboratory has developed a desktop machining microfactory [7, 43] consisting of machine tools such as a lathe, a milling machine, a press machine, and assembly machines such as a transfer arm and a two-fingered hand. This portable microfactory has external dimensions of 625 x 490 x 380 mm^3 and weighs 34 kg (its main body weight is 23 kg). It uses three miniature CCD cameras mounted on each machine tool, which display the image of a machined section on a 5.8-inch LCD monitor. This factory can produce and assemble miniature machine parts. Only two joysticks and one push button are used to operate the equipment. The authors of this concept have reported some features of their microfactory, such as:

1. A marked reduction in the consumption of drive energy and system environment maintenance energy (air conditioning, illumination, etc.);
2. A decreased inertial force, facilitating the control of prompt enhancement, speed and positioning precision increase;
3. Miniaturization and integration, increasing the degree of freedom of product design, and facilitating the modification of system layout.

The overall equipment sizes in this microfactory vary from $30 \times 30 \times 30$ mm^3 to $150 \times 150 \times 150$ mm^3. With this equipment, different samples of microdetails and microdevices were produced. For example, ball bearings with an external diameter of 0.9 mm [44] were made. Thin needles with a diameter of 50 μm and a length of 600 μm [43] were successfully machined out of brass.

The idea of microfactory creation is supported also in other countries. In Switzerland, the details of precision motion control and microhandling principles for future microfactories have been worked out [45]. Using electrochemical discharge machining, this group has machined 500 μm diameter shapes such as pentagons or screws from conductive cylindrical samples. Friedrich and Vasile [38] have shown that a micromilling process permits the machining of 3D walls of 10–15 μm thickness. Thus, in principle, it is possible to manufacture 3D microdetails using conventional mechanical treatment processes. It has also been shown that it is possible to realise such manufacturing in desktop microfactories. The advantages and drawbacks of this approach are discussed below. We attempt to show how to avoid the main drawbacks of microfactory equipment as well as to describe two developed prototypes of micromachine tools.

The microfactory must contain automatic machine tools, automatic assembly devices, robots for feeding machine tools, assembly devices, devices for quality inspection, waste elimination systems, transport systems, inspection of tool condition, a system for tool replacement, and so on. The Japanese microfactory [7] contains many components from this list. Microfactory creation is intended to dramatically save energy, space, and resources at production plants by reducing the size of production machines to that comparable with the products [46]. Below, we give an analysis of these advantages of microfactories. The analysis shows that these advantages are not sufficient for effective promotion of microfactories to the market and discovers other features of microfactories that could promote their use. The main drawback of the microfactory prototype developed in Japan is the relatively high cost of devices and components. For effective use of microfactories, it is necessary to create them on the basis of low-cost components as well as to solve problems of low-cost automation of these microfactories to reduce the labor cost.

Let us analyze the advantages of microfactory creation in the context of component costs. A conventional machine tool requires approximately 0.5 kW to produce both macrodetails and microdetails. If the cost of 1 kWh is \$0.10, this machine tool can work 10,000 hours to consume energy costing \$500. A micromachine tool consumes less than 0.05 kW [47]. The cost of energy for the micromachine tool for the same period will be less than \$50. The profit in this case will be almost \$500, which seems large. But if the cost of the micromachine tool components is \$5,000 (our rough estimate), this profit is only 10%. This profit is insufficient for replacement of ordinary machine tools by micromachine tools in manufacturing processes.

Let us consider space saving from the replacement of conventional machine tools by micromachine tools. Conventional machine tools can take up 5 m^2 (including service area). One micromachine tool from the Japanese microfactory occupies 0.09 m^2 including service area. This means that practically all the area of the conventional machine tool will be saved. If the rent of one square meter is \$50 per year, the rent payment for the conventional machine tool will be \$250 per year. For five years, the savings will be \$1,250, which could serve as the stimulus for the replacement of conventional machine tools but is insufficient for fast promotion and implementation of microequipment if its components have a high cost.

Regarding material consumption, we can roughly estimate this by the weight of the machine tools. Conventional machine tools for production of small details weigh approximately 150 kg. The weight of a micromachine tool is negligible in comparison with a conventional machine tool. If the mean cost of materials is \$3 per kilogram, then the savings from materials will be \$450. This is also insignificant for the fast promotion of micromachine tools.

This rough analysis shows that, for more effective promotion of microfactories to the market, it is necessary to seek ways of decreasing the costs of microequipment components. As a rule, super-precise components are too expensive for use in microequipment (for example, an interferometer may cost some thousands of dollars) [45]. Less expensive devices, such as precise linear encoders, may cost \$100–200, but this is also expensive for microequipment. In this context, it is interesting to investigate the possibility of creating microequipment using low-cost, imprecise components and to obtain the required relative precision of treated micropieces from the decrease of machine tool sizes and the use of adaptive treatment algorithms.

In this chapter, we describe two prototypes of micromachine tools having components that cost less than \$100. In MET, the precision of machine tools decreases due to low-cost components. To restore precision, it is necessary to use different methods. The theoretical analysis of these methods and experimental investigations are presented in this paper.

8.2 General Rules for Scaling Down Micromechanical Device Parameters

The changing of mechanical and electrical parameters of microdevices during the scaling down process was investigated by W. S. N. Trimmer [48] and other authors [41]. Trimmer considered two microdevices of the same design but different sizes. The scaling factor of linear sizes is S. According to his results, magnetic forces can decrease as S^4, S^3, or S^2 depending on the other parameters' scaling. S^4 corresponds to an electromagnetic system having the same current density in the windings of both devices; S^3 corresponds to a system that includes a permanent magnet; and S^2 corresponds to the case of equal operating temperature of the windings. The electrostatic forces scale as S^2 for the constant electric field and S when the electric field scales as $S^{-0.5}$. Trimmer analyzed parameters such as accelaration and time of part movements. Depending on the scaling force, these parameters could scale as:

$$a = \begin{bmatrix} S^{-2} \\ S^{-1} \\ S^{0} \\ S^{1} \end{bmatrix}, \tag{8.1}$$

$$t = \begin{bmatrix} S^{1,5} \\ S^{1} \\ S^{0,5} \\ S^{0} \end{bmatrix}, \tag{8.2}$$

where a is the acceleration and t is the time of movement.

Trimmer also analyzed power generation and dissipation. If the force scales as S^2, then the power per unit volume scales as S^{-1}. On the basis of this rule, Trimmer concludes that electromagnetic motors will not be effective in microscale and should be replaced by other types of micromotors. We will consider this problem below.

Ishihara H., Arai F., and Fukuda T. [41] have analyzed the scaling of different forces. They gave the following results (Table 8.1). In defining the scaling effect for an electrostatic force, H. Ishihara et al. assume that the voltage between the electrodes does not change after the device size changes. In reality, the voltage applied to the electrodes, as a rule, is reduced approximately linearly with a reduction of device size. Taking this change into account, the scaling effect for electrostatic force will be S^2, not S^0.

In defining the scaling effect for an inertial force, the authors of the article accept the time of the displacement of some mass as constant. In real devices, the linear velocity of device components is constant more often than not. We will discuss the reasons for this phenomenon later. Let us consider the uniformly accelerated movement of mass m over distance d. Let the initial velocity equal v_1 and final velocity equal v_2. Then, from the law of conservation of energy, the inertial force is:

$$F_i d = m\frac{v_2^2}{2} - m\frac{v_1^2}{2}. \tag{8.3}$$

Table 8.1 Scaling effect for different forces

Kind of Force	Symbol	Equation	Scaling Effect	
Electromagnetic Force	F_{magc}	$\frac{B}{2\mu}S_m$	S^2	μ - permeability; B - magnetic field density; S_m- area of cross section of coil; S - scaling factor
Electrostatic Force	F_{static}	$\frac{\varepsilon S_m}{2}\frac{V^2}{d^2}$	S^0	ε - permittivity; V - applied voltage; S_m - surface area; d - gap between electrodes; S - scaling factor
Thermal Expansion Force	F_{ther}	$ES_m\frac{\Delta L(T)}{L}$	S^2	E - Young's Modulus; L - length; ΔL - strain; T - temprature; S -scaling factor
Piezoelectric Force	F_{piezo}	$ES_m\frac{\Delta L(E_e)}{L}$	S^2	E - Young's Modulus; L - length; ΔL - strain; T - temprature; S -scaling factor
Inertial Force	F_i	$m\frac{\partial^2 x}{\partial t^2}$	S^4	m - mass; t - time; x - displacement; S- scaling factor
Viscosity Force	F_v	$c\frac{S_m}{L}\frac{\partial x}{\partial t}$	S^2	c - viscosity coefficient; x -displacement; S_m - surface area; L - length; t - time; S - scaling factor
Elastic Force	F_e	$ES_m\frac{\Delta L}{L}$	S^2	E - Young's Modulus; S_m - cross section area; ΔL - strain; L - length; S - scaling factor

Hence:

$$F_i = \frac{m(v_2^2 - v_1^2)}{2d}. \tag{8.4}$$

Since

$$m_A = S^3 m_B, \tag{8.5}$$

$$d_A = Sb_B, \tag{8.6}$$

we have the equation:

$$F_{i_A} = F_{i_B} S^2, \tag{8.7}$$

where m_A is the moving mass in the machine tool A; m_B is the moving mass in the machine tool B; d_A is the distance mass A is moving; d_B is the distance mass B is moving; F_{i_A} is the inertial force in the machine tool A; and F_{i_A} is the inertial force in the machine tool B. Equation (8.7) shows that the scaling effect for the inertial force under linear displacement equals S^2, not S^4.

To estimate the influence of the scaling effect on micromachine tool precision, it is necessary to include an analysis of the elastic and thermal deformations of the micromachine tool components. We have performed this work [49, 50] and present the results in the next chapter.

8.3 The Analysis of Micromachine Tool Errors

8.3.1 *Thermal Expansion*

Let us consider machine tool *A* and machine tool *B*, which have the same design and are made of the same materials. Let each component size of machine tool *B* be *S* times smaller than the corresponding component size of machine tool *A*. The thermal expansion of a component having length *L* can be presented as:

$$\Delta L = \alpha \cdot L \cdot \Delta T, \tag{8.8}$$

where ΔL is the thermal deformation, α is the thermal expansion coefficient, and ΔT is the difference in temperature. The term α is the same in machine tools *A* and *B* because it depends only on the component material. The length of a component *B* is *S*-times smaller than the length of a component *A*:

$$L_A = S \cdot L_B. \tag{8.9}$$

The temperature difference, ΔT, consists of two parts: ΔT_i, the internal temperature difference, and ΔT_e, the external temperature difference. Machine tool *B* demands smaller room volume than machine tool *A*. As it is easier to maintain a constant temperature in the smaller volume than in the larger one, we can write:

$$\Delta T_{e_A} \geq \Delta T_{e_B}. \tag{8.10}$$

To compare the internal temperature difference, let us consider a simplified scheme of the origin of temperature difference shown in Fig. 8.2. Let the temperature T_2 of the heat sink (HS) be constant and the energy produced by the heat generator be *W*. The temperature difference can be obtained from:

$$\Delta T = T_1 - T_2 = \frac{W \cdot L_{HT}}{\lambda \cdot S_{HT}}, \tag{8.11}$$

where L_{HT} is the length of the heat transfer (HT) element, S_{HT} is the square area of the HT element, and λ is the heat transfer coefficient. The term λ is the same in machine tools *A* and *B*. For L_{HT} and S_{HT} we have:

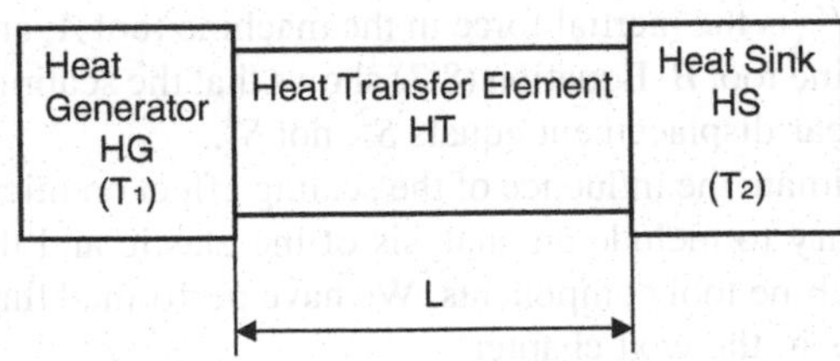

Fig. 8.2 Scheme of the origin of temperature difference

$$L_{HT_A} = S \cdot L_{HT_B}, \tag{8.12}$$

$$S_{HT_A} = S^2 \cdot S_{HT_B}. \tag{8.13}$$

We estimate the term W as follows:

$$W = w \cdot V_{H\phi}, \tag{8.14}$$

where $V_{H\phi}$ is the heat generator volume, and w is the volumetric energy produced by one unit of heat generator volume.

We consider two main cases. In the first case, the term w is the same in machine tools A and B, leading to

$$W_A = S^3 \cdot W_B, \tag{8.15}$$

because $V_A = S^3 \cdot V_B$.

On the basis of (8.11) – (8.13) and (8.15) we obtain:

$$\Delta T_A = \frac{W_A \cdot L_{HT_A}}{\lambda \cdot S_{HT_A}} = \frac{S^3 \cdot W_B \cdot S \cdot L_{HT_B}}{\lambda \cdot S^2 \cdot S_{HT_B}} = S^2 \cdot \Delta T_B. \tag{8.16}$$

In the second case, the volumetric energy increases with a decrease of the machine tool size:

$$w_A = \frac{w_B}{S}. \tag{8.17}$$

Therefore:

$$W_A = S^2 \cdot W_B, \tag{8.18}$$

and

$$\Delta T_A = \frac{W_A \cdot L_{HT_A}}{\lambda \cdot S_{HT_A}} = \frac{S^2 \cdot W_B \cdot S \cdot L_{HT_B}}{\lambda \cdot S^2 \cdot S_{HT_B}} = S \cdot \Delta T_B. \tag{8.19}$$

Both cases show that the internal temperature difference in machine tool B is smaller than that in machine tool A.

Returning to equation (8.8), we can obtain:

$$\Delta L_A = \alpha \cdot L_A \cdot \Delta T_A \geq \alpha \cdot S \cdot L_B \cdot \Delta T_B = S \cdot \Delta L_B, \tag{8.20}$$

which means that the thermal deformations decrease linearly (or faster) with a decrease of the machine tool sizes.

8.3.2 Rigidity

The rigidity r is determined from:

$$r = F_z/\Delta z, \tag{8.21}$$

where F_z is the force that affects the object, and Δz is the deformation of this object in the force direction.

We will consider several examples.

8.3.2.1 Compression (or extension) of the bar

The force F_z compresses the bar with the section q, producing the deformation Δz (Fig. 8.3). In this case we have:

$$\Delta_z = (F_z \cdot L)/(q \cdot E), \tag{8.22}$$

where E is the modulus of material elasticity of the bar. From equations (8.21) and (8.22) we obtain:

$$r = F_z/\Delta_z = (q \cdot E) / L. \tag{8.23}$$

In accordance with the initial supposition for devices A and B, we have:

$$L_A = S \cdot L_B, \tag{8.24}$$

$$q_A = S^2 \cdot q_B, \tag{8.25}$$

$$E_A = E_B. \tag{8.26}$$

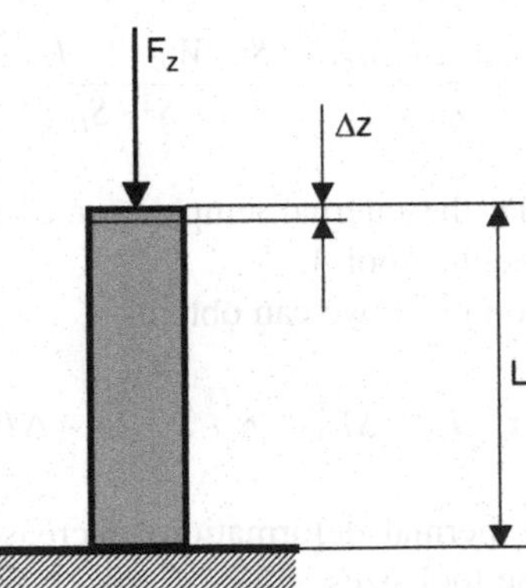

Fig. 8.3 The bar compression

From these equations we have:

$$r_A = \frac{q_A \cdot E_A}{L_A} = \frac{S^2 \cdot q_B \cdot E_B}{S \cdot L_B} = S \cdot r_B. \tag{8.27}$$

Equation (8.27) shows that the rigidity of bar *A* is *S* times greater than the rigidity of bar *B*.

8.3.2.2 Bending of the bar. Case 1

The bar of length *L* is loaded at one end by the force F_z. The other end is fixed to the wall (Fig. 8.4). The bar has a constant section with the moment of inertia *I*. In this case, the deflection Δz of the bar end is:

$$\Delta z = \frac{F_Z \cdot L^3}{3 \cdot E \cdot I}, \tag{8.28}$$

and the rigidity *r* is:

$$r = F_Z/\Delta z = \frac{3 \cdot E \cdot I}{L^3}. \tag{8.29}$$

We have:

$$\begin{aligned} E_A &= E_B, \\ L_A &= S \cdot L_B, \\ I_A &= S^4 \cdot I_B. \end{aligned} \tag{8.30}$$

Substitution into equation (8.29) gives:

$$r_A = \frac{3 \cdot E_A \cdot I_A}{L_A^3} = \frac{3 \cdot E_B \cdot S^4 \cdot I_B}{S^3 \cdot L_B^3} = S \cdot r_B. \tag{8.31}$$

We can conclude from equation (8.31) that a size reduction by *S* times gives a rigidity reduction by *S* times.

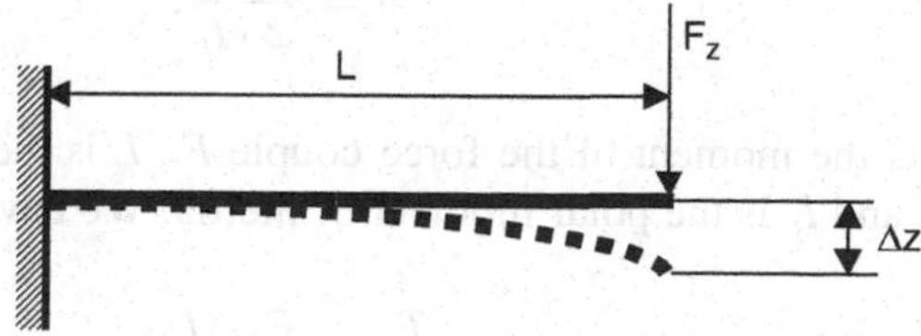

Fig. 8.4 Bending of the console bar

Fig. 8.5 Bending of the bar

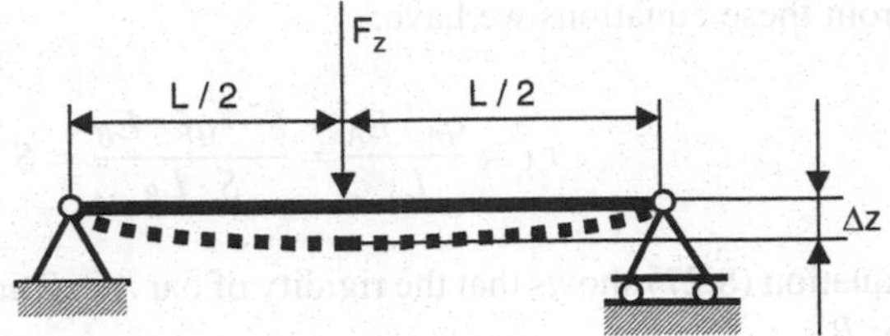

8.3.2.3 Bending of the bar. Case 2

The bar of length L is loaded at the center by the force F_z. Both ends of the bar are fixed to the supports (Fig. 8.5). The bar section has the moment of inertia I. For this case, the deflection Δz is:

$$\Delta z = \frac{F_Z \cdot L^3}{48 \cdot E \cdot I}, \tag{8.32}$$

and the rigidity r is:

$$r = Fz/\Delta z = \frac{48 \cdot E \cdot I}{L^3}. \tag{8.33}$$

Substitution of equation (8.30) into equation (8.33) gives:

$$r_A = \frac{48 \cdot E_A \cdot I_A}{L_A^3} = \frac{48 \cdot E_B \cdot S^4 \cdot I_B}{S^3 \cdot L_B^3} = S * r_B \cdot \tag{8.34}$$

We can conclude from equation (8.34) that a size reduction by S times gives a rigidity reduction by S times.

8.3.2.4 Torsion of the bar

The bar of length L is fixed to the wall (Fig. 8.6). The end of this bar is fixed perpendicularly to another bar of length L_1. Both ends of bar L_1 are loaded by the forces F_z that form the force couple. This couple turns the bar L that has a polar moment of inertia I_P and produces a displacement Δz of the ends of bar L_1. Bar L_1 is considered absolutely rigid. The angle θ of the torsion of bar L can be calculated using the formula:

$$\theta = \frac{T \cdot L}{\varphi \cdot I_p}, \tag{8.35}$$

where T is the moment of the force couple F_z, L is the bar length, φ is the shear modulus, and I_p is the polar moment of inertia. We have:

$$T = F_z \cdot L_1, \tag{8.36}$$

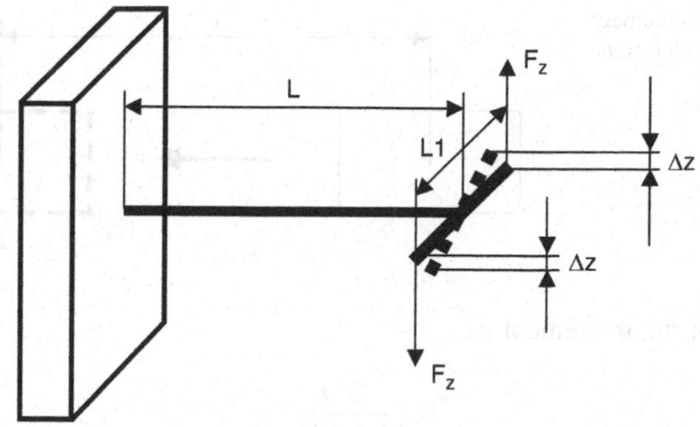

Fig. 8.6 Torsion of the bar

$$\Delta z = \theta\left(\frac{L_1}{2}\right) = \frac{F_z \cdot L_1 \cdot L}{\varphi \cdot I_p} \cdot \frac{L_1}{2} = \frac{F_z \cdot L_1^2 \cdot L}{2 \cdot \varphi \cdot I_p} \tag{8.37}$$

$$r = \frac{F_z}{\Delta z} = \frac{2 \cdot \varphi \cdot I_p}{L_1^2 \cdot L} \tag{8.38}$$

As in previous cases we have:

$$\begin{aligned} L_A &= S \cdot L_B, \\ L_{IA} &= S \cdot L_{IB}, \\ \varphi_A &= \varphi_B, \end{aligned} \tag{8.39}$$

$I_{PA} = S^4 \cdot I_{PB}$

Substitution of equation (8.39) into equation (8.38) gives:

$$r_A = \frac{2 \cdot \varphi_A \cdot I_{PA}}{L_{1A}^2 \cdot L_A} = \frac{2 \cdot \varphi_B \cdot \ S^4 \ \cdot I_{PB}}{S^2 \cdot \ L_{1B}^2 \cdot S \cdot L_B} = S \cdot r_B \cdot \tag{8.40}$$

The examples (8.3.2.1), (8.3.2.2), (8.3.2.3), (8.3.2.4) show that the rigidity of device *A* is always greater than the rigidity of device *B* by *S* times. In other words, the rigidity decreases linearly with the microdevice size diminution.

8.3.3 *Forces of Inertia*

8.3.3.1 Force of inertia. Linear movement with uniform acceleration

A component of the mass *m* moves from point 1 to point 2 with uniform acceleration (Fig. 8.7). The distance between points 1 and 2 is *L*. The initial velocity is V_1, and the terminal velocity is V_2.

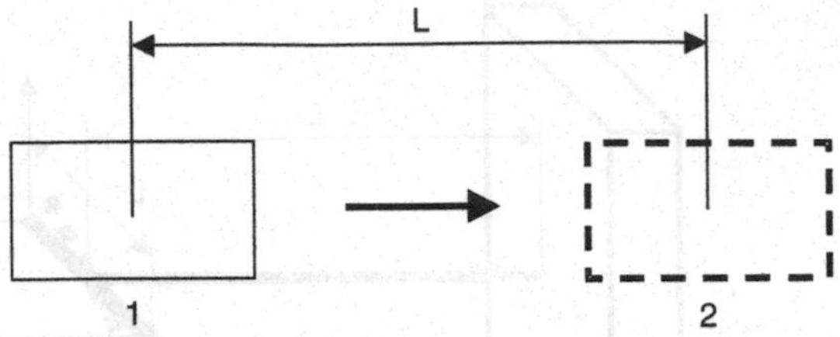

Fig. 8.7 Linear movement with uniform acceleration

The time of the movement is:

$$t = \frac{2 \cdot L}{V_1 + V_2}. \tag{8.41}$$

The acceleration is:

$$W = \frac{V_2 - V_1}{t} = \frac{(V_2 - V_1)(V_2 + V_1)}{2 \cdot L} = \frac{V_2^2 - V_1^2}{2 \cdot L} \tag{8.42}$$

The inertial force is:

$$F = m \cdot W = \frac{m \cdot \left(V_2^2 - V_1^2\right)}{2 \cdot L} \tag{8.43}$$

With the condition $V = const$:

$$F_A = \frac{m_A \cdot \left(V_{2A}^2 - V_{1A}^2\right)}{2L_A} = \frac{S^3 \cdot m_B \cdot \left(V_{2B}^2 - V_{1B}^2\right)}{2 \cdot S \cdot L_B} = S^2 \cdot F_B \tag{8.44}$$

With the condition $t = const$:

$$F_A = \frac{m_A \cdot \left(V_{2A}^2 - V_{1A}^2\right)}{2L_A} = \frac{S^3 \cdot m_B \cdot \left(S^2 \cdot V_{2B}^2 - S^2 \cdot V_{1B}^2\right)}{2 \cdot S \cdot L_B} = S^4 \cdot F_B \tag{8.45}$$

8.3.3.2 Centrifugal force

A disc of mass m rotates around the axis with angular velocity $\bar{\omega}$ (Fig. 8.8). The disc is fixed on the axis with eccentricity ΔR. The centrifugal force is:

$$F = m \cdot \bar{\omega}^2 \cdot \Delta R = \frac{m \cdot V^2}{\Delta R}, \tag{8.46}$$

where V is the linear circular velocity of the disc center. With condition $V = const$ we will have:

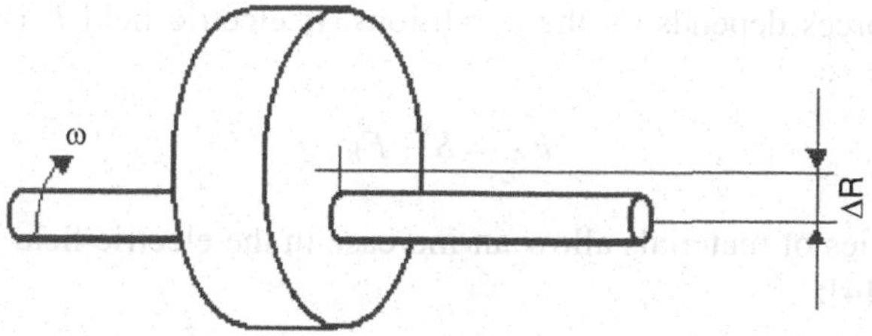

Fig. 8.8 Rotation of the disc

$$F_A = \frac{m_A \cdot V_A^2}{\Delta R_A} = \frac{S^3 \cdot m_B \cdot V_B^2}{S \cdot \Delta R_B} = S^2 \cdot F_B. \tag{8.47}$$

With condition $t = const$ we will have:

$$F_A = m_A \cdot \bar{\omega}_A^2 \cdot \Delta R_A = S^3 \cdot m_B \cdot \bar{\omega}_B^2 \cdot S \cdot \Delta R_B = S^4 \cdot F_B \tag{8.48}$$

8.3.4 Magnetic Forces

W. S. N. Trimmer [4, 48] investigated the scaling of magnetic, electrostatic forces and power generation and dissipation. For magnetic forces, there are different scaling conditions. If, for example, devices A and B are electromagnetic motors and current density in the windings of these motors is constant, the forces in motors A and B scale as:

$$F_A = S^4 \cdot F_B. \tag{8.49}$$

If these motors have permanent magnet rotors, the force scaling will be:

$$F_A = S^3 \cdot F_B. \tag{8.50}$$

However, a small motor has better conditions for refrigeration; therefore, it is possible to increase the current density in motor B until the temperature of its windings is equal to the temperature of motor A's windings. In this case, Trimmer obtained the formula:

$$F_A = S^2 \cdot F_B. \tag{8.51}$$

8.3.5 Electrostatic Forces

In conventional mechanics, electrostatic actuators are not used in the motors because the electrostatic forces are weak. In [4, 48] it is shown that scaling of electrostatic forces depends on the conditions. If electric field E is constant, the force is:

$$F_A = S^2 \cdot F_B. \tag{8.52}$$

The properties of materials allow an increase in the electric field in accordance with equation [4]:

$$E_A = \frac{E_B}{\sqrt{S}}. \tag{8.53}$$

In this case:

$$F_A = S \cdot F_B. \tag{8.54}$$

Trimmer analyzed the possibility of using magnetic and electrostatic forces to create micromotors. The scaling of electrostatic forces (equation (8.54)) allows an increase of the volumetric power, diminishing the micromotor size. But experience with MEMS micromotors demonstrates that, in the micrometric range, the electrostatic electromotors still do not have sufficient power for use in micromachine tools and micromanipulators. In this sense, they cannot compete with micromotors based, for example, on piezoelectric forces.

The magnetic forces, which scale in accordance with equation (8.51), in principle, can be used for micromotor creation, but the micromotor efficiency η decreases in accordance with the equation:

$$\eta_A = S \cdot \eta_B. \tag{8.55}$$

If these motors are to be used in microfactories, the consumption of volumetric energy will grow linearly with the diminution of the micromachine tool sizes, making it necessary to search for other forces for micromotor creation. In MEMS, many different principles of motor operation were tested, such as piezoelectric motors, thermal motors, motors based on shape memory alloys (SMAs), pneumatic motors, and hydraulic motors. The piezoelectric motors have many problems with the wearing of their components. The thermal motors and SMA motors have low efficiency, η. The pneumatic and hydraulic motors have great volumetric power and can have high efficiency, so it is interesting to consider their scaling.

8.3.6 *Viscosity and Velocity of Flow*

8.3.6.1 Pneumatic and hydraulic forces

Let us consider two hydraulic or pneumatic cylinders A and B with pistons. The size of cylinder A is S times larger than cylinder B.

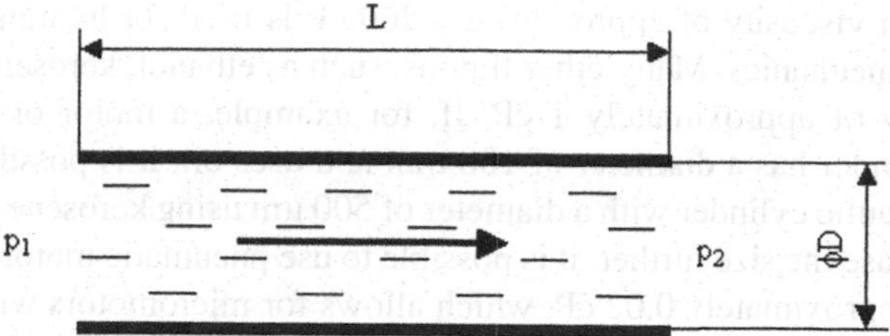

Fig. 8.9 Internal friction in the liquid or gas

Condition 1. The pressure difference, Δp, and the viscosity, μ, of the liquid or gas is constant. The pressure force on the piston is:

$$F_A = \Delta p_A \cdot A_A = \Delta p_B \cdot S^2 \cdot A_B = S^2 \cdot F_B, \tag{8.56}$$

where A is the area of the cylinder, and Δp is the pressure difference on the piston. The velocity V of the piston could be limited by the internal friction of the liquid or gas in the channel that feeds them to the cylinder (Fig. 8.9). To estimate the velocity limit, it is necessary to calculate the flow rate, Q, using the equation:

$$Q = \frac{\pi \cdot D^4}{128 \cdot \mu \cdot L} \cdot (p_1 - p_2), \tag{8.57}$$

where D and L are the diameter and the length of the channel, p_1 and p_2 are the inlet pressure and the exit pressure of the channel, and μ is the viscosity. From equation (8.57) we have:

$$Q_A = \frac{\pi \cdot D_A^4}{128 \cdot \mu_A \cdot L_A} \cdot (p_1 - p_2) = \frac{\pi \cdot S^4 \cdot D_B^4}{128 \cdot \mu_B \cdot S \cdot L_B} \cdot (p_1 - p_2) = S^3 \cdot Q_B \tag{8.58}$$

From formula (8.58) we conclude that the flow rate decreases with the cube of the device size. In this case, the velocity of channel flow can be calculated as:

$$V_A = \frac{Q_A \cdot 4}{\pi \cdot D_A^2} = \frac{S^3 \cdot Q_B \cdot 4}{\pi \cdot S^2 \cdot D_B^2} = S \cdot V_B. \tag{8.59}$$

Equation (8.59) is equivalent to the condition $t = const.$

Condition 2. We mentioned that a more advanced condition is $V = const.$ In order to obtain this condition, it is necessary to decrease the viscosity of the liquid or gas in device B in accordance with the equation:

$$\mu_A = S \cdot \mu_B. \tag{8.60}$$

An oil with viscosity of approximately 200 cP is used for hydraulic motors in conventional mechanics. Many other liquids, such as ethanol, kerosene, and water, have viscosity of approximately 1 cP. If, for example, a motor or conventional hydraulic cylinder has a diameter of 100 mm and uses oil, it is possible to make a motor or hydraulic cylinder with a diameter of 500 μm using kerosene or ethanol. In order to decrease the size further, it is possible to use pneumatic motors. Gases have viscosity of approximately 0.02 cP, which allows for micromotors with a diameter of 10 μm. So, using hydraulic or pneumatic micromotors, it is possible to develop micromachine tools and micromanipulators in the range needed for micromechanical devices.

8.3.6.2 Forces of surface tension (capillary forces)

A round bar with diameter d touches a base (Fig. 8.10). The water steam is condensed in the contact area. The superficial tension produces the capillary force F that tightens the bar to the base:

$$F = \sigma \cdot \pi \cdot d. \tag{8.61}$$

This force scaling is:

$$F_A = \sigma_A \cdot \pi \cdot d_A = \sigma_B \cdot \pi \cdot S \cdot d_B = S \cdot F_B. \tag{8.62}$$

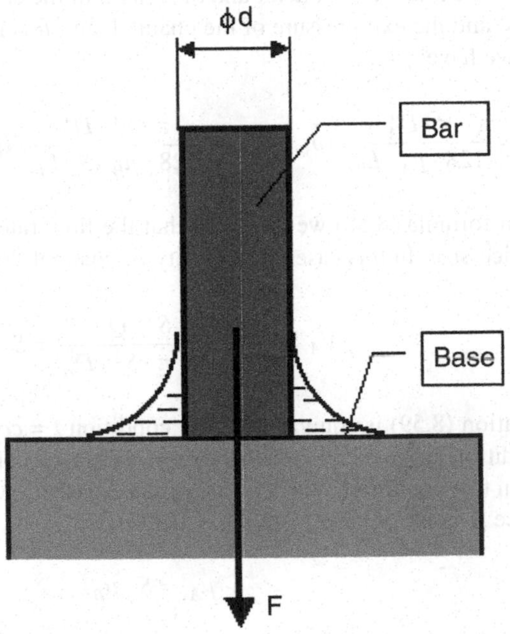

Fig. 8.10 Capillary forces

It is possible to estimate the order of magnitude of the capillary force. For d = 1 μm (10^{-6} m) and σ = 7.10^{-2} N/m we have:

$$F = \sigma \cdot \pi \cdot d = 7 \cdot 10^{-2} \cdot 3.14 \cdot 10^{-6} = 2.2 \cdot 10^{-7} N. \tag{8.63}$$

For comparison, we calculated the hydraulic force, F_h, that affects a piston of the same diameter 1 μm, if the pressure difference is $\Delta p = 10^7$ Pa.

$$F_h = \frac{\Delta p \cdot \pi \cdot d^2}{4} = \frac{10^7 \cdot 3.14 \cdot 10^{-12}}{4} = 7.85 \cdot 10^{-6} N \tag{8.64}$$

The capillary force is much lower than the hydraulic force.

8.3.7 Mass Forces

To calculate the mass forces of a component of devices A and B, we divide the component into the elementary parallelepipeds of sizes $dx * dy * dz$ in such a way that each parallelepiped of device A has a corresponding parallelepiped of device B. In this case:

$$\begin{aligned} dx_A &= S \cdot dx_B \\ dy_A &= S \cdot dy_B \\ dz_A &= S \cdot dz_B \end{aligned} \tag{8.65}$$

The component mass can be defined as:

$$m = \iiint \rho(x, y, z) \cdot dx \cdot dy \cdot dz, \tag{8.66}$$

where ρ (x, y, z) is the material density in coordinate (x, y, z). All parallelepipeds of component A are made from the same materials as in component B; thus:

$$\rho_A(x_A, y_A, z_Z) = \rho_B(x_B, y_B, z_B) \tag{8.67}$$

From equations (8.65), (8.66), and (8.67) we obtain:

$$\begin{aligned} m_A &= \iiint \rho_A(x_A, y_A, z_A) \cdot dx_A \cdot dy_A \cdot dz_A \\ &= \iiint \rho_B(x_B, y_B, z_B) \cdot S \cdot dx_B \cdot S \cdot dy_B \cdot S \cdot dz_B = S^3 \cdot m_B \end{aligned} \tag{8.68}$$

8.3.8 Forces of Cutting

The cutter produces chips during the cutting process. The chip formation is shown in Fig. 8.11. In this figure, a short cutter cuts raw material and produces chips

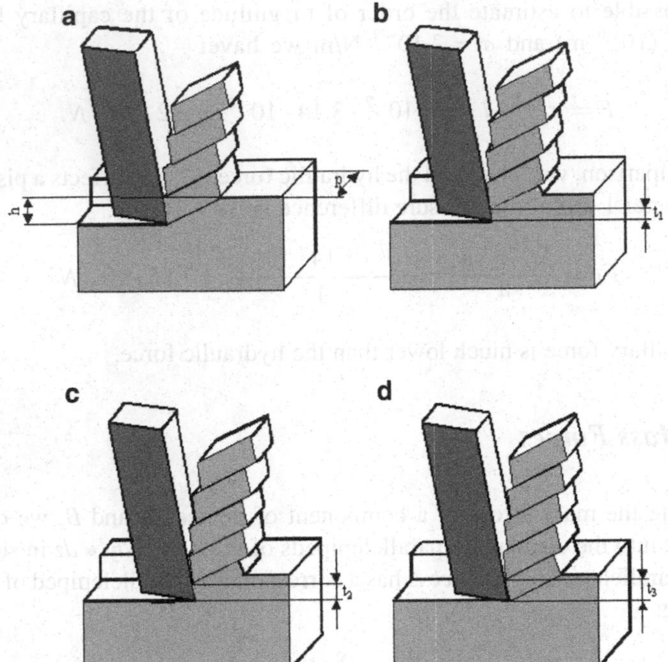

Fig. 8.11 Scheme of chip production

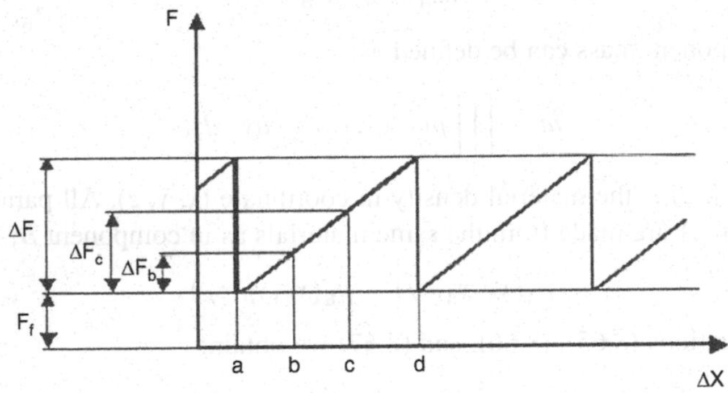

Fig. 8.12 The force of chip formation

consisting of small pieces of the raw material. In Fig. 8.11, the small lower piece adheres to the raw material, and the force that affects the cutter is the force of friction, F_f (Fig. 8.12).

When the cutter moves forward (Fig. 8.11b), it crushes raw material. The force is increased by the value:

$$\Delta F_b = C \cdot \sigma_T \cdot t_1 \cdot p, \tag{8.69}$$

where C is a constant, σ_T is the specific tension of the compression of raw material, t_1 is the height of the squashed area (Fig. 8.11b), and p is the cutting depth (Fig. 8.11a). When the cutter moves forward, the additional force increases (Fig. 8.11c) and reaches the maximum value ΔF in the position presented in Fig. 8.11d:

$$\Delta F = C \cdot \sigma_T \cdot t_3 \cdot p. \tag{8.70}$$

We can describe the scaling of this force as:

$$\Delta F_A = C \cdot \sigma_T \cdot t_{3A} \cdot p_A = C \cdot \sigma_T \cdot S \cdot t_{3B} \cdot S \cdot p_B = S^2 \cdot \Delta F_B. \tag{8.71}$$

8.3.9 Elastic Deformations

Elastic deformations are one of the more important causes of error in mechanical device production with the help of machine tools. From formula (8.21) we obtain the elastic deformation:

$$\Delta z = \frac{F}{r}, \tag{8.72}$$

where r is the machine tool rigidity, and F is the force. From equations (8.27), (8.31), (8.34), and (8.40) we have:

$$r_A = S \cdot r_B. \tag{8.73}$$

For the many cases presented in equations (8.44), (8.47), (8.51), (8.52), (8.56), and (8.71), the force is:

$$F_A = S^2 \cdot F_B. \tag{8.74}$$

For these cases, the deformation scaling is:

$$\Delta z_A = \frac{F_A}{r_A} = \frac{S^2 \cdot F_B}{S \cdot r_B} = S \cdot \Delta z_B. \tag{8.75}$$

In other cases [equations (8.45), (8.48), (8.49)] we have:

$$F_A = S^4 \cdot F_B, \tag{8.76}$$

and

$$\Delta z_A = \frac{F_A}{r_A} = \frac{S^4 \cdot F_B}{S \cdot r_B} = S^3 \cdot \Delta z_B. \tag{8.77}$$

For case (8.50) we have:

$$\Delta z_A = \frac{F_A}{r_A} = \frac{S^3 \cdot F_B}{S \cdot r_B} = S^2 \cdot \Delta z_B. \tag{8.78}$$

We can conclude that, in the majority of the considered examples, the errors caused by elastic deformation decrease linearly (or faster) with the diminution of the micromachine tool size. The exceptions are presented in the cases of equations (8.54) and (8.62), where:

$$F_A = S \cdot F_B \tag{8.79}$$

and

$$\Delta z_A = \frac{F_A}{r_A} = \frac{S \cdot F_B}{S \cdot r_B} = \Delta z_B. \tag{8.80}$$

These cases correspond to the electrostatic force or capillary force. These forces are of an order of magnitude lower than other forces until they reach a size of 1 μm, which is why it is possible not to take into account their influence.

8.3.10 Vibrations

Vibrations contribute a sufficiently large percentage of machine tool errors. To calculate the deformations from vibrations, let us consider a disc fixed on a rotating shaft (Fig. 8.13). The center point of the disc is displaced relative to the center point of the shaft. The displacement is ε. The inertial force F_i of the disc can be presented as:

$$F_i = m \cdot \omega^2 \cdot \varepsilon, \tag{8.81}$$

where m is the mass of the disc, and ω is the angular speed of disc rotation. The shaft can be considered as a console beam loaded at the end by the inertial force, F_i (Fig. 8.14).

The deflection, Δz, from beam bending is:

$$\Delta z = C \cdot \frac{L^3 \cdot F_i}{E \cdot D^4}, \tag{8.82}$$

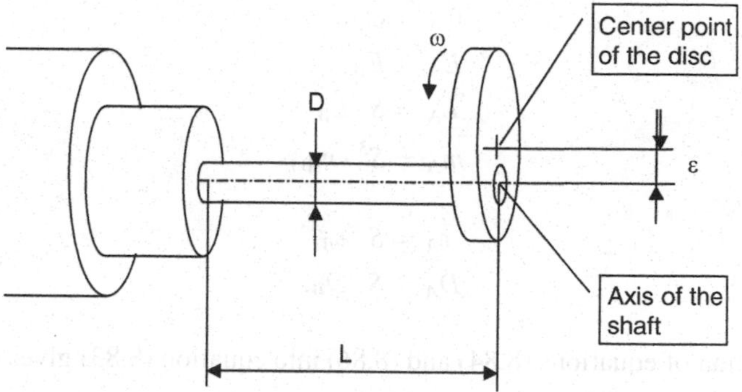

Fig. 8.13 The disc fixed on the rotating shaft

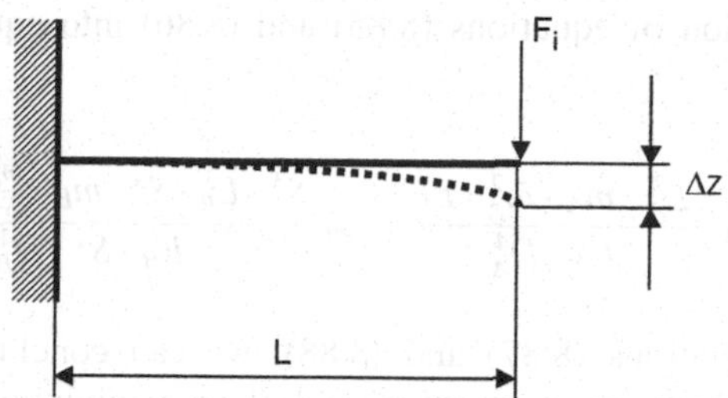

Fig. 8.14 The console beam loaded at the end by the inertial force Fi

where C is a constant, E is the elastic module, which depends on shaft material, L is the shaft length, D is the shaft diameter, and F_i is the inertial force. From (8.81) and (8.82) we obtain:

$$\Delta z = C \cdot \frac{L^3 \cdot m \cdot \bar{\omega}^2 \cdot \varepsilon}{E \cdot D^4} \tag{8.83}$$

For comparison, we consider two cases of machine tools A and B. In the first case

$$\bar{\omega}_{\mathrm{A}} = \bar{\omega}_{\mathrm{B}} \tag{8.84}$$

and in the second case

$$\bar{\omega}_{\mathrm{A}} = \frac{\bar{\omega}_{\mathrm{B}}}{S}. \tag{8.85}$$

In both cases:

$$
\begin{aligned}
C_A &= C_B, \\
E_A &= E_B, \\
L_A &= S \cdot L_B, \\
m_A &= S^3 \cdot m_B, \\
\\
\varepsilon_A &= S \cdot \varepsilon_B, \\
D_A &= S \cdot D_B.
\end{aligned}
\tag{8.86}
$$

Substitution of equations (8.84) and (8.86) into equation (8.83) gives for the first case:

$$
\Delta z_A = C_A \cdot \frac{L_A^3 \cdot m_A \cdot \bar{\omega}_A^2 \cdot \varepsilon_A}{E_A \cdot D_A^4} = C_B \frac{S^3 \cdot L_B^3 \cdot S^3 \cdot m_B \cdot \bar{\omega}_B^2 \cdot S \cdot \varepsilon_B}{E_B \cdot S^4 \cdot D_B^4} = S^3 \cdot \Delta z_B. \tag{8.87}
$$

Substitution of equations (8.85) and (8.86) into equation (8.83) gives for the second case:

$$
\Delta z_A = C_A \cdot \frac{L_A^3 \cdot m_A \cdot \bar{\omega}_A^2 \cdot \varepsilon_A}{E_A \cdot D_A^4} = C_B \frac{S^3 \cdot L_B^3 \cdot S^3 \cdot m_B \cdot \frac{\bar{\omega}_B^2}{S^2} \cdot S \cdot \varepsilon_B}{E_B \cdot S^4 \cdot D_B^4} = S \cdot \Delta z_B. \tag{8.88}
$$

From equations (8.87) and (8.88), we can conclude that the inertial forces produce deflections in machine tool B as a minimum S times smaller than in machine tool A. Inertial forces present the main component of the vibration process. Another important component is the quality factor of the resonant oscillation. There are data [4] showing that the quality factor also decreases with a decrease in the size of mechanical devices.

A similar analysis of other error sources shows that errors decrease at least linearly with a decrease in machine tool size. So, it is a good idea to decrease machine tool size down to that comparable with workpieces. However, modern technology does not permit the production of micromachine tools having an overall size of some millimetres. That is why, in the Japanese microfactory project, the size of the machine tools is three to ten times larger than the initially proposed size [46]. To solve this problem, we propose making the microequipment through sequential generations of machine tools and manipulators of smaller and smaller sizes.

8.4 The First Prototype of the Micromachine Tool

The developed micromachine tool prototype is used for turning, milling, drilling, and grinding [5 – 8]. To change the type of work, it is necessary to change tools and programs. The prototype was tested in the turning mode. In its design, we used the criteria of low cost and the possibility of scaling down the micromachine tool.

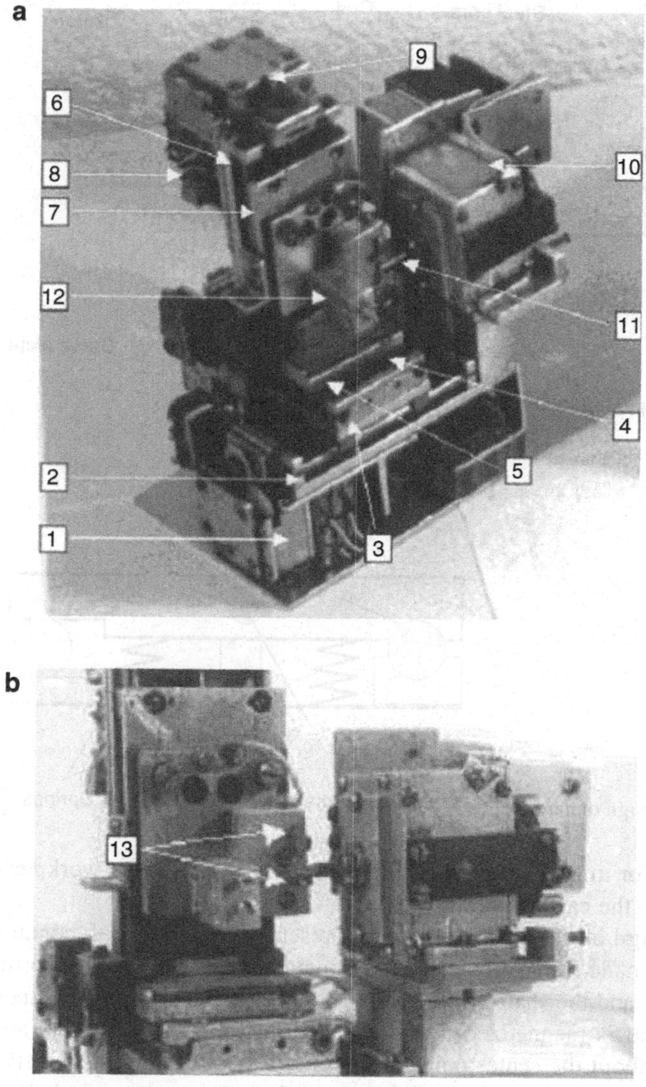

Fig. 8.15a, b The first micromachine tool prototype

The first prototype is shown in Fig. 8.15*a*, *b*. On the base (1), the guides (2), (4), (6) for three carriages (3), (5), (7) are installed by the sequential scheme, i.e., each subsequent guide is installed on the previous carriage to provide translation movements along the *X*, *Y*, and *Z* axes. The spindle case (10) with spindle (11) is also installed on the base. The drives for the carriages and for the spindle use stepping motors (8) with gearboxes (9). The spindle has a chuck to grip the workpiece for turning, a drill for drilling, or a mill for milling. There is a gripper (12) with a cutter, and there are two parallel metal pins (13) for measurement of the turned workpiece

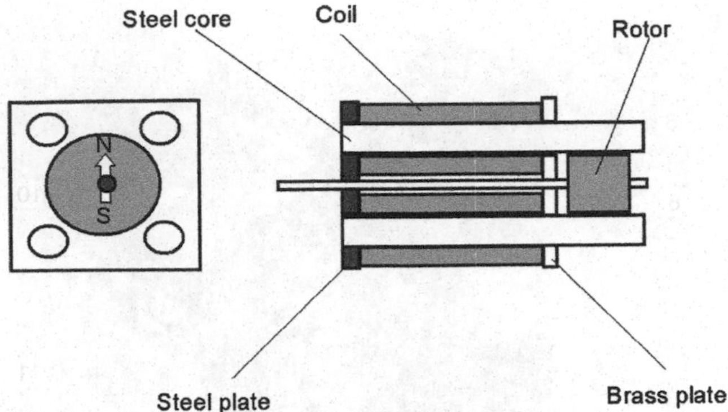

Fig. 8.16 Stepping motor

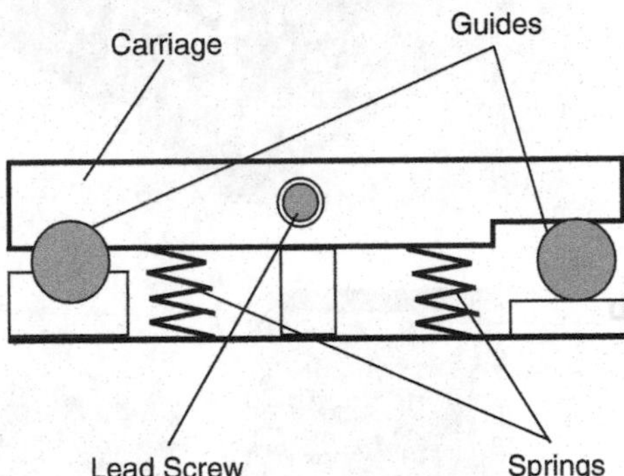

Fig. 8.17 Design of carriage

diameter. For milling and drilling, the special gripper for the workpiece should be installed on the carriage (7).

The design of the stepping motor is presented in Fig. 8.16. It contains four coils with cores made from soft steel, a brass plate, and a steel plate supporting the cores of the coils and the shaft of the rotor. The rotor is made from permanent magnetic material and is magnetized in the radial direction. The coil pairs are connected crosswise about the center point, so the motor has four input contacts. The motor makes four steps per revolution and can be forced to make eight half-steps. In our experiments, the stepping motor was used in the mode of high-speed rotation. We used gearboxes to decrease the rotation speed. The main advantage of this motor is the simplicity of manufacture and assembly. In addition, this motor does not contain details much smaller than the motor, which is very important for scaling down.

The guides for the carriages were made as round bars, which simplifies the production, assembly, and scaling down of micromachine tools. The carriages (Fig. 8.17) were made as simple as possible, each containing a spring to eliminate the spaces between the carriage and the guides. Such spaces could appear due to variations of guide diameter and thermal expansion of the carriage.

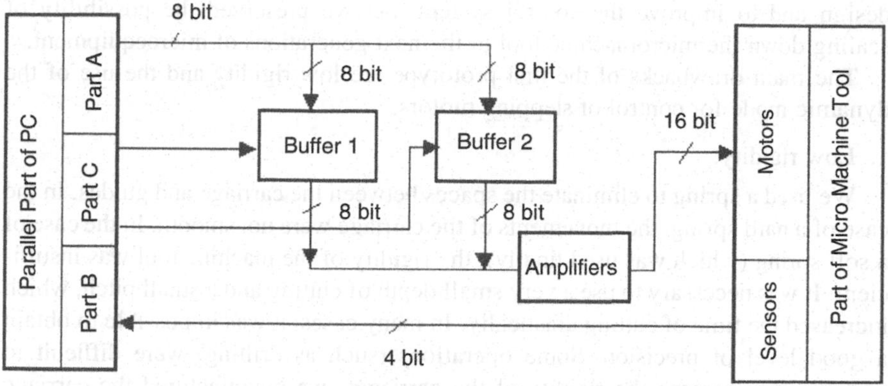

Fig. 8.18 Control system scheme of the first prototype

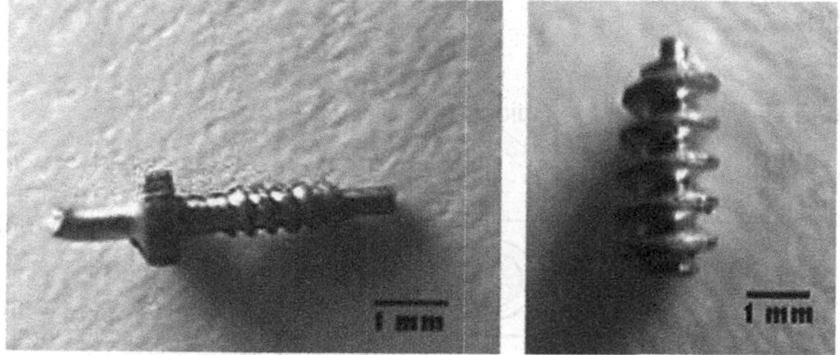

Fig. 8.19 Devices manufactured with the first micromachine tool prototype

All other parts of the first prototype of the micromachine tool were also made according to the criteria of low cost and scaling down. Figure 8.18 shows a schematic diagram of the control system of the first prototype. This system was designed to use the possibilities of a personal computer (PC) and to minimize the external equipment. The external equipment contains only two 8-bit buffers and a 16-channel amplifier of direct current. The port of the micromachine tool contains only connectors and transformers of sensor contact signals of the voltage levels that are compatible with the input PC port. Devices manufactured with the first micromachine tool prototype are shown in Fig. 8.19.

8.5 The Second Prototype

The main advantages of the first prototype are the simplicity of its design and the easy manufacture of its components, along with a very simple control system. But the first prototype had some drawbacks, which we tried to eliminate in the second prototype [8]. It was necessary to increase slightly the complexity of the prototype design and to improve the control system, but we preserved the possibility of scaling down the micromachine tool in the next generations of microequipment.

The main drawbacks of the first prototype are low rigidity and the use of the dynamic mode for control of stepping motors.

1. Low rigidity.

We used a spring to eliminate the spaces between the carriage and guides. In the case of a hard spring, the movements of the carriage were not smooth. In the case of a soft spring (which was used finally), the rigidity of the machine tool was insufficient. It was necessary to use a very small depth of cutting and a small pitch, which increased the time of cutting drastically. In many cases, it was impossible to obtain a good level of precision. Some operations (such as drilling) were difficult to perform. To increase the rigidity of the carriages, we manufactured the carriage guides more precisely and changed the springs to a hard support. To compensate for differences in the distance between the two guides, ball-based supports were used (see Fig. 8.20).

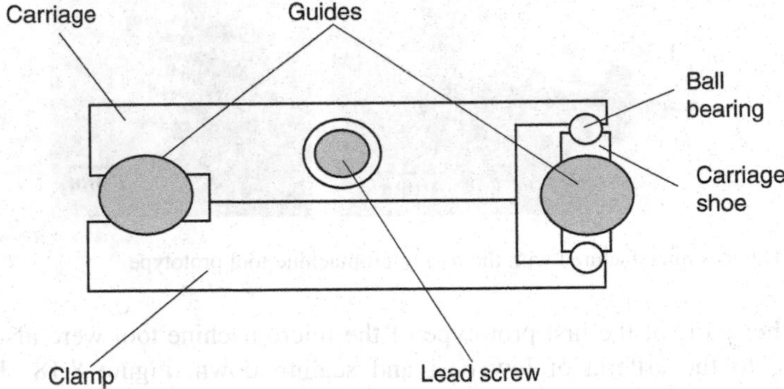

Fig. 8.20 Design of second carriage prototype

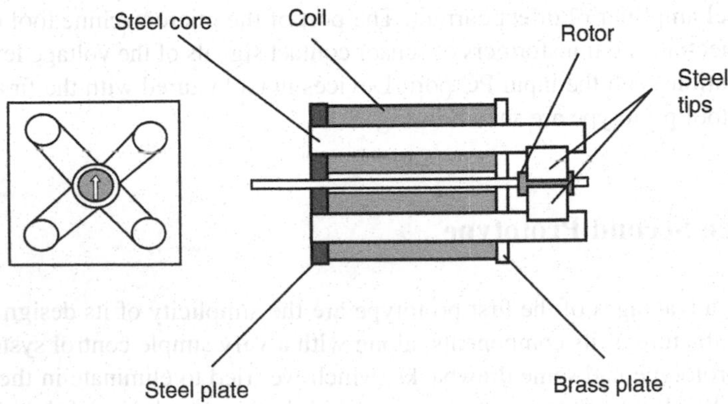

Fig. 8.21 Second prototype stepping motor

Fig. 8.22 The second micromachine tool prototype

2. Dynamic model of stepping motors.

Many problems were connected with the dynamic mode of the stepping motors, which permitted us to work only with DOS. Windows programs could not maintain real-time control of the stepping motor in the dynamic mode. To preserve a relatively high rotation speed without the dynamic mode, it was necessary to decrease the rotor's inertial moment. For this purpose, the rotor diameters in the second prototype motors were made almost three times smaller than those of the first prototype. The second prototype stepping motor is shown in Fig. 8.21. Finally, we obtained the rotation speed of the new stepping motors in quasi-static mode, and this was even larger than the dynamic mode speed of the first prototype motors.

Figure 8.22 shows the second prototype, which we have been exploiting for approximately forty months for experimental work and student training. We have disassembled the prototype for wearing inspection and cleaning on two occasions. The wearing was observed to be negligible. One stepping motor was replaced.

8.6 The Second Micromachine Tool Prototype Characterization

The second prototype of the micromachine tool has size $130 \times 160 \times 85\ \text{mm}^3$ and is controlled by a PC. The *X* and *Z* axes have 20 mm of displacement, and the *Y* axis has 35 mm of displacement; all axes have the same configuration. The theoretical resolution is 1.87 μm per the motor step.

In general, almost all micromechanical devices are hard to evaluate. The characteristics of the prototype can be obtained by direct or indirect methods. The direct method is based on mathematical equations that describe the machine tool errors. In machine tools, there are many different error sources associated with the design of the mechanical parts and their characteristics. One of these error sources is the geometrical error in the guideway systems [51]. These errors are associated with the squareness between all the axes, the straightness of each axis, the rotation of the axes, and the nonlinear behavior of the lead screw. The decomposition of the errors in each axis is shown in the following equations:

$$\delta_x = f_1(x) + y\cos\alpha_{xy} + z\cos\alpha_{xz}, \quad (8.89)$$

$$\delta_y = f_2(y) + x\cos\alpha_{yx} + z\cos\alpha_{yz}, \quad (8.90)$$

$$\delta_z = f_3(z) + x\cos\alpha_{zx} + y\cos\alpha_{zy}, \quad (8.91)$$

$$\alpha_{xy} = \varphi_1(x, y), \quad (8.94)$$

$$\alpha_{xz} = \varphi_2(x, z), \quad (8.93)$$

$$\alpha_{yz} = \varphi_2(y, z), \quad (8.94)$$

where δ represents the total error associated with each axis, f is the error function of the lead screw, and φ is the error function of straightness, rotation, and squareness.

At present, the characterization of this prototype has been made by an indirect method, which means by the inspection of its own products (manufactured pieces). We have used the following parameters for characterization of the second prototype:

- *Positional characteristics*. Analysis of the axes' resolution (linear displacement for each motor step). We also found backlash errors.
- *Geometrical inspection*. Geometrical analysis of the parameters of the pieces produced on our microequipment and a comparison of these with design parameters of pieces provides accuracy characteristics.

Tests of the second micromachine prototype were made with the production of two test pieces with approximately known geometric characteristics. Using statistical analysis, we determined the required parameters. For this analysis, 20 samples of each test piece were employed. Figure 8.23 shows the samples of the test pieces.

8.6.1 Positional Characteristics

One of the test pieces was a cylinder with the three marks 1, 2, and 3 (Fig. 8.23). This was used to determine the positional characteristics [8, 52]. The measured

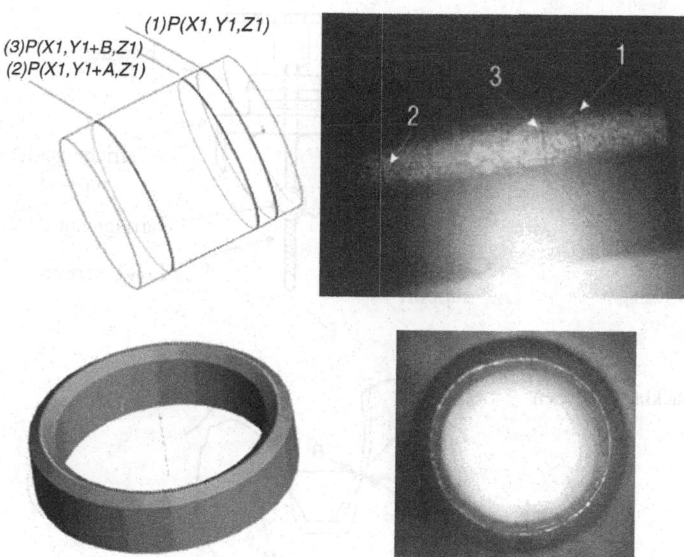

Fig. 8.23 Test pieces employed for the second prototype evaluation

length was the distance between marks 1 and 2. This magnitude corresponds to the programmed displacement of 1,050 motor steps. The results of the measurements are $X_{1-2} = 1978$ μm, $\sigma = 6.07$ μm. The standard deviation σ gives information over 62.8 % of the samples.

In order to obtain a better sample range, it is necessary to triplicate the value of this result to get a sampling of 99.3%, i.e., $3\sigma = 18.2$ μm. Because of this, the error of this sample is $Error = \pm$ 18.2 μm. This is the average error, and it can be interpreted as the error of the *Y*-axis displacement because all measurements were made in different positions on the axis. The displacement resolution for each step is $X_{DPSM} = 1.88$ μm, $\sigma = 0.005$ μm. In this case, the standard deviation is helpful only to give us an idea about the behavior of the resolution compared with the theoretical expectations for the deviation, but it is not a result that we can trust because the error is lower than the equipment accuracy. This standard deviation might be an estimation of the random error.

In order to define the backlash presented on the *Y*-axis, we measured the distance between marks 1–3 of the testing pieces (Fig. 8.23). The average backlash of the *Y*-axis is $B_g = 345.3$ μm, $\sigma = 7.8$ μm. There are different backlash error sources, the first of which is connected with the gearbox. The second and most important source is produced by the transmission and the carriage on each axis. The leading screw and the nut that is coupled to the carriage on each axis produce the third source (Fig. 8.24) [52 – 55]. It is possible to estimate the backlash produced by the gearbox using the measurements of gear and shaft parameters. The measurement of transmission and leading screw parameters is much more complicated. Therefore, we calculate the gearbox backlash B_g and then obtain transmission and lead screw backlash B_a as a difference between real backlash B_t and gearbox backlash B_g.

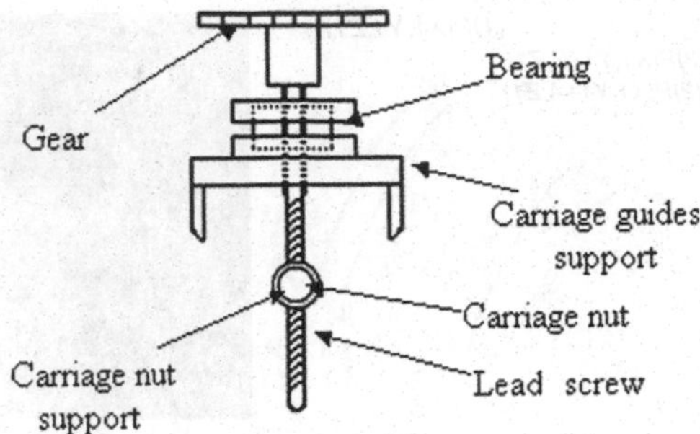

Fig. 8.24 Sources of backlash

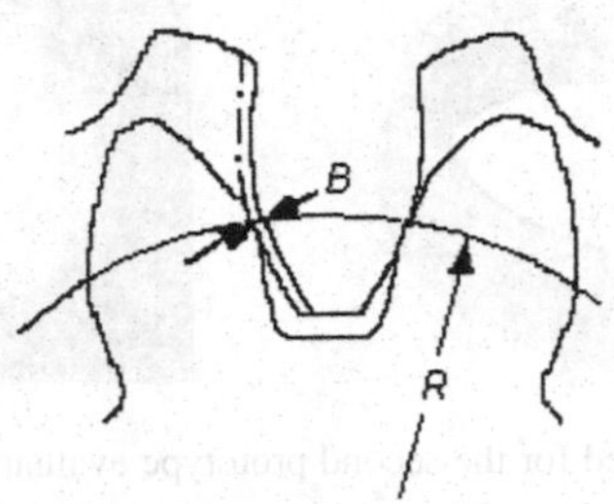

Fig. 8.25 Backlash between two gears

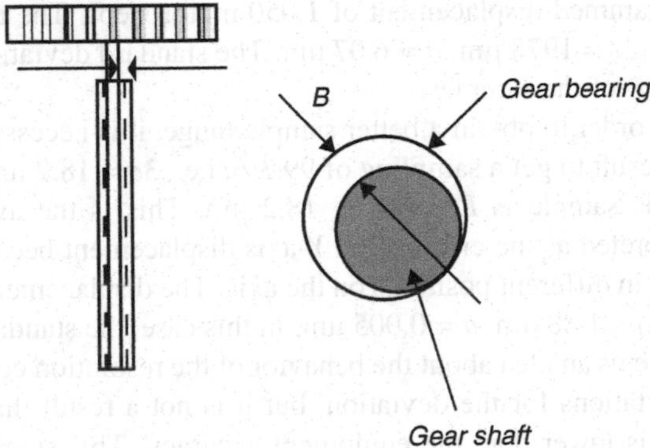

Fig. 8.26 Backlash from the space between the gear shaft and the gear hold

Three important components of gear train backlash are:

1. The space between the gear teeth at the operating pitch circle (Fig. 8.25);
2. The space between the gear shaft and the gear hold (Fig. 8.26); and
3. The deformation of the console gear shaft (Fig. 8.27).

In the first case, to define the backlash between the gear teeth, we use the equation:

$$BA = \frac{B}{R}, \tag{8.95}$$

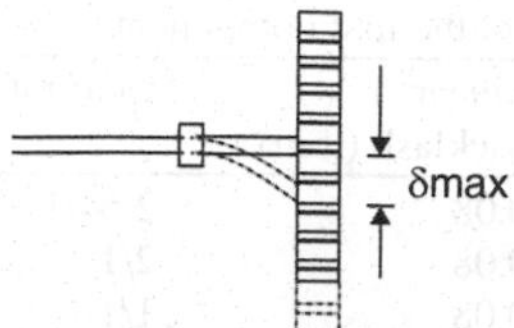

Fig. 8.27 Backlash from the deformation of the console gear shaft

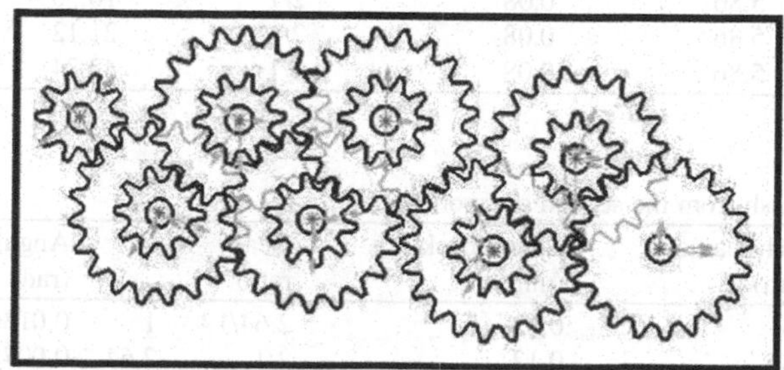

Fig. 8.28 Gear train in the micromachine tool axis

where BA is the angular backlash of the gear train element, B is the linear backlash, and R is the pitch radius of the gear. Then the backlash of the gear train is:

$$BA^j = \sum_{i=1}^{N} \frac{B_i}{R_i v_i}, \tag{8.96}$$

where B_i is the linear backlash, v_i is the relative rotation speed of the ith shaft, R_i is the pitch radius of the ith gear, N is the number of gear pairs, and j is the number of backlash components. Figure 8.28 shows the gear train in the micromachine tool axis. The backlash results of the gear train are shown in Table 8.2.

The backlash of the gear train is $BA^1 = 0.0216$ *rad* or $BA^1 = 1.2427°$. To obtain the backlash from the second component (the space between the gear shaft and the gear hold), we use equation (8.96). The results are shown in Table 8.3.

The backlash is $BA^2 = 0.0325$ rad or $BA^2 = 1.8641°$. The backlash from the deformation of the console gear shaft was calculated with the gear train working with the maximum load. The equation represents the maximum deformation of the console gear shaft. (We suppose that the lead screw is fixed in the bearing and that the axial force between the last gears generates the load applied on the lead screw.)

$$\delta_{\max} = \frac{F_L l^3}{3EI}, \tag{8.97}$$

where F_L is the load, l is the length of the console part of the shaft, E is the modulus of elasticity, and I is the moment of inertia of the shaft cross section. F_L is obtained from the torque of the motor and the gear ratio.

Table 8.2 Backlash of gear train from the first component

Gear number	Pitch radius (mm)	Linear backlash (mm)	Gear ratio	v	Angular backlash (rad)
1	8.81	0.08	2.64/1	1	0.0091
3	5.86	0.08	2/1	2.64	0.0052
5	5.86	0.08	1/1	5.28	0.0026
6	5.86	0.08	2/1	5.28	0.0026
8	5.86	0.08	2/1	10.56	0.0013
10	5.86	0.08	2/1	21.12	0.0006
12	5.86	0.08	2/1	42.24	0.0003

Table 8.3 Backlash from the second component

Gear number	Pitch diameter (mm)	Linear backlash (mm)	Gear ratio	v	Angular backlash (rad)
1	8.81	0.12	2.64/1	1	0.0136
3	5.86	0.12	2/1	2.64	0.0078
5	5.86	0.12	1/1	5.28	0.0039
6	5.86	0.12	2/1	5.28	0.0039
8	5.86	0.12	2/1	10.56	0.0019
10	5.86	0.12	2/1	21.12	0.001
12	5.86	0.12	2/1	42.24	0.0005

$$P_m = \bar{\omega}_m T_m, \tag{8.98}$$

$$P_m \cong P_L, \tag{8.99}$$

$$T_m = \frac{1}{a} T_L, \tag{8.100}$$

where P_m is the motor power, P_L is the lead screw power, $\bar{\omega}_m$ is the motor angular velocity, T_m is the motor torque, T_L is the lead screw torque, and a is the gear ratio.

To obtain the lead screw torque, we use equation (8.100) where the motor torque is 0.003 Nm and the gear ratio is 84.48; $T_L = 84.48 \times 0.003 = 0.25344$ Nm. The load on the lead screw is

$$F_L = \frac{T_L}{R} = \frac{0.25344}{0.0088} = 28.8\,N, \tag{8.101}$$

where R is the pitch radius. The moment of inertia of area I is obtained by

$$I = \frac{\pi r^2}{4} = \frac{\pi \times 0.0015^4}{4} = 3.976 \times 10^{-12}\text{m}^4, \tag{8.102}$$

where r is the radius of the console shaft cross section. The length of lead screw console part l is 0.01 m, and the module elasticity E is 207×10^9 N/m^{-2}. If we substitute these results in equation (8.97), we obtain

$$\delta_{\max} = \frac{28.8 \times .01^3}{3 \times 207 \times 10^9 \times 3.976 \times 10^{-12}} = 0.0000117\,m = 11.7\ \mu\text{m}.$$

The backlash from the console deformation is:

$$BA^3 = \frac{\delta_{\max}}{R} = \frac{11.7}{8800} = 0.0013,$$
$$BA^3 = 0.075^\circ. \tag{8.103}$$

The backlash of the gear train is:

$$BA = BA^1 + BA^2 + BA^3 = 3.088^\circ \tag{8.104}$$

The linear displacement of the carriage per revolution of the last gear is 635 μm; thus, the total backlash produced by the gear train is

$$B_g = \frac{BA}{360} \times d = \frac{3.088}{360} \times 635 = 5.4\ \mu\text{m}, \tag{8.105}$$

where d is the linear displacement of the carriage per revolution of the last gear.

The total experimental backlash for each axis of the micromachine tool is 345.3 μm. Other components of the backlash are from the spaces between the lead screw and the supporting bearing, between the lead screw and the carriage nut, and between the carriage nut and the carriage nut support (Fig. 8.24). We call this backlash B_a. The total backlash is:

$$B_t = B_a + B_g \tag{8.106}$$

From equations (8.105) and (8.106) we have: $B_a = B_t - B_g = 339.9\ \mu\text{m}$.

For backlash compensation we use correction algorithms, which minimize the influence of the backlash on the accuracy of the machine tool. For this purpose, all the work is performed using unidirectional movements. When it is necessary to turn the carriage back, a special program accounts for the backlash in order to compensate for it.

The other important parameter of the machine tool is the repeatability of carriage movements. We define (1) the repeatability of measurements for each axis without return to the home position and (2) the repeatability with return to the home position (all axes). In the test, we use two balls; one ball is fixed at a point within the workspace of the micromachine tool and the other is fixed on the tool support (Fig. 8.29).

In the first test, the ball fixed on the tool support moves and makes contact with the other ball in the Y direction. The positions along the Y-axis are read and saved in a file. After this, the tool support goes back along the same axis to a random point and returns to make contact again, and we save the coordinates of this point. This process is repeated as many times as necessary. In our tests, we made

Fig. 8.29 Ball location for the repeatability test

20 independent measurements. For the X and Z axes we repeat the same procedure. Finally, the repeatability for each axis is defined as the difference between the maximum and the minimum obtained for each axis. The results of the test are as follows: $X_{max} = 8{,}378$; $X_{min} = 8{,}371$; $Y_{max} = 7{,}093$; $Y_{min} = 7{,}083$; $Z_{max} = 4{,}597$; and $Z_{min} = 4{,}591$.

The repeatability in each axis is:

$$R_x = N_{xmax} - N_{xmin} = 8,378 - 8,371 = 7 \text{ (steps)} \tag{8.107}$$

$$R_y = N_{ymax} - N_{ymin} = 7,093 - 7,083 = 10 \text{ (steps)} \tag{8.108}$$

$$R_z = N_{zmax} - N_{zmin} = 4,597 - 4,591 = 6 \text{ (steps)} \tag{8.109}$$

To know the repeatability in micrometers, it is necessary to multiply the results by 1.88 (displacement of each axis per step): $R_x = 13.16$ µm; $R_y = 18.8$ µm; and $R_z = 11.28$ µm.

For the second test, a special computer program was written by Alberto Caballero. The ball fixed to the tool support is moved by 3,000 steps in the Y axis and 3,000 steps in the Z axis from the home position, and makes contact with another ball in the X direction. The position on the X-axis is read and saved in a file. After this, the machine returns to the home position. This process is repeated as many times as necessary. The results of this test are as follows: $X_{max} = 2{,}082$; $X_{min} = 2{,}067$; $Y = 3{,}000$; $Z = 3{,}000$. The home position repeatability is: $R = N_{max} - N_{min} = 2{,}072 - 2{,}067 = 15$ (steps) or $R = 1.88 \times 15 = 28.2$ µm. In this case, the home position repeatability is larger than the repeatability without the home position. For this reason, we have used the home position procedure only at the start of the workpiece treatment. The errors of the home position procedure were compensated by intermediate measurements of the workpiece sizes.

8.6.2 Geometric Inspection

Microrings were employed for the geometric inspection. These rings were designed for a micromechanic filter (one of the applications of the proposed technology). The solid model and sample of the microring are shown in Fig. 8.23. Table 8.4 contains the measurements of the microrings. The corresponding measurements are presented in Fig. 8.30. The measurement results are shown in Table 8.5.

Errors of the produced pieces depend on the machine tool, the instruments, and the details machined. These errors allow for the characterization of the micromachining center. For this reason, we use this indirect method to characterize the micromachine tool. The results obtained are satisfactory because the difference between the practical and theoretical results is sufficiently low. The comparison is shown in Table 8.6.

Table 8.4 The measurements of microrings for geometrical analysis

Measurement name	Description
M	Exterior diameter of the microring base.
S	Distance between the diameters (interior and exterior of the microring base)
N	Diameter corresponding to the intersection of conic surfaces
P	Distance between the conic surface intersection and the interior diameter of the cylindrical part of the microring
Q	Height of the cylindrical part of the microring

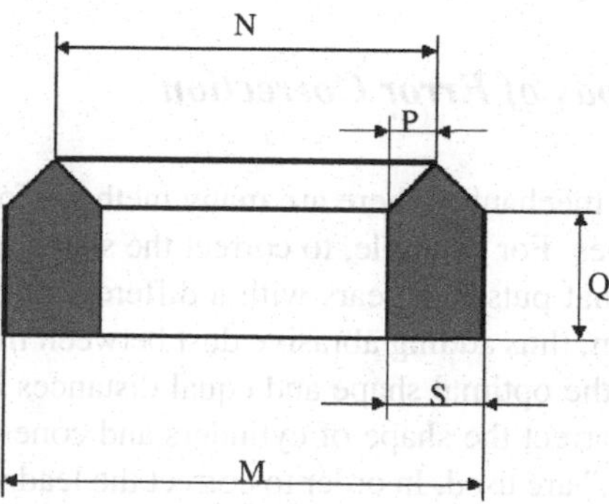

Fig. 8.30 The microring measurements

Table 8.5 Measurement results

Machining type	Average measurement (μm)	Average error ±(μm)
Cylinder machining (M)	1258.45	20.11
Drilling (S)	205.47	38.28
Cone machining (N)	981.30	62.19
Cone machining (P)	64.63	37.00
Cone machining (Q)	476.79	34.82

Table 8.6 A comparison of the theoretical and the practical results

Characteristics	Theoretical value (μm)	Measured value (μm)
Displacement resolution	1.875	1.88
Y-axis accuracy	±18	±18
Backlash	300	345.27
Machining accuracy	–	±62

8.7 Errors that Do Not Decrease Automatically

In section 8.3, we showed that a decrease in microequipment size automatically decreases the errors of micromachine tools and micromanipulators. However, not all errors follow this rule. For example, if we produce a lead screw for microlathe *B* using microlathe *A* of the previous generation and the lead screw of microlathe *A* has step error Δx_A, then the lead screw of microlathe *B* can have the same error:

$$\Delta x_B = \Delta x_A. \tag{8.110}$$

The same properties can appear in the deviations of the circular shape of the chuck components. Also, the shape errors of the gear teeth could be conserved during scaling down. To diminish these errors, it is necessary to propose additional methods that could correct errors. Below, we propose methods to solve this problem.

8.7.1 *Methods of Error Correction*

In conventional mechanics, there are many methods for autocorrecting shapes and sizes of the pieces. For example, to correct the shape and the pitch of gear teeth, a device is used that puts two gears with a different number of teeth in contact and then rotates them, thus adding abrasive dust between the teeth. During this process, all teeth obtain the optimal shape and equal distances between them.

In order to correct the shape of cylinders and cones, methods of "super finish," such as "honing," are used. In order to correct the lead screw shape, a long, split nut (Fig. 8.31) can be used. In Fig. 8.31, the nut (2) is put on the lead screw (3) and is tightened with the screw (4). The screw (3) periodically rotates in different directions, and the nut moves along the screw (3) to the left and to the right. Through the nozzle (1), a suspension of liquid with abrasive dust is added to the area of nut and lead screw contact. After many movements, the shape of the lead screw thread and the thread pitch are automatically equalized.

These and other methods derived from conventional mechanics permit a reduction of the exceptional errors. The disadvantage of these methods is low productivity

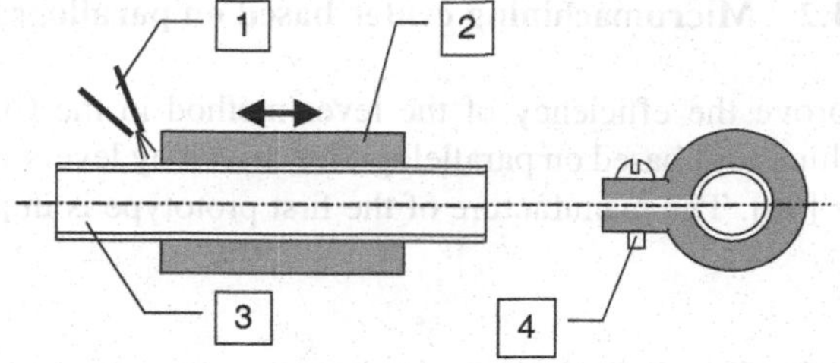

Fig. 8.31 Autocorrection of the lead screw

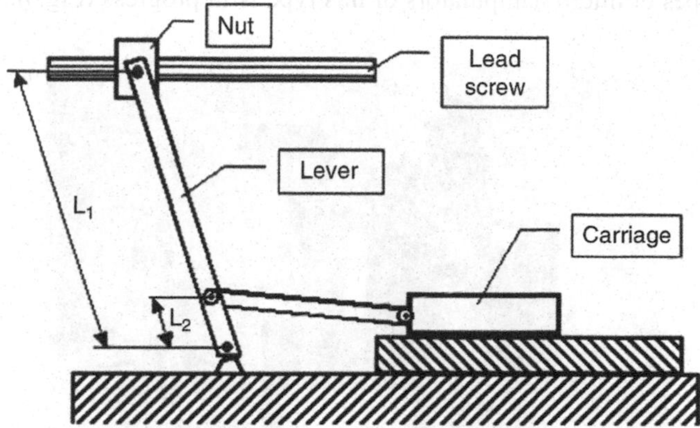

Fig. 8.32 The scheme of the "lever" method

and large wearing of the tools. For this reason, we also propose other methods that allow a reduction of the exceptional errors.

8.7.1.1 The method of "lever."

The scheme of the "lever" method is presented in Fig. 8.32. The carriage moves linearly on a table with a lever. One end of this lever is connected to the table, and the other end is connected to the nut that is moving with the lead screw. All errors of the lead screw decrease by the factor L_1 / L_2 when the movement is transferred from the nut to the carriage. This scheme allows a reduction of the lead screw errors when we produce the following generation components. For example, if we attach a tool holder with a cutter to the carriage and want to make a lead screw for a smaller micromachine tool, this smaller lead screw will have relative errors (errors divided by screw size) equal to the relative errors of the large lead screw. This method can be used not only in machine tools but also in micromanipulators.

8.7.1.2 Micromachining center based on parallograms

To prove the efficiency of the lever method in the CCADET, UNAM, a micromachine tool based on parallelograms driven by levers was developed by Leopoldo Ruiz [56]. The manufacture of the first prototype is in progress (Fig. 8.33).

8.7.1.3 Parallel micromanipulators

The lever method will be proved in parallel micromanipulators. The manufacture of two prototypes of micromanipulators of this type is in progress (Fig. 8.34).

Fig. 8.33 Microcenter based on parallelograms (in progress)

Fig. 8.34 Parallel micromanipulator

8.8 Adaptive Algorithms

It is possible to increase micromachine tool precision using adaptive algorithms. In this section, we describe two relevant adaptive algorithms. The first, which is described below, is based on a sensor of electrical contact, and it allowed us to produce a brass pin of 50 μm diameter (Fig. 8.35). The same result was obtained in the Japanese project.

8.8.1 Adaptive Algorithms Based on a Contact Sensor

Let us consider the problem of turning a fine pin with a microlathe [52–55]. The errors of our micromachine tool do not allow us to obtain the needed precision. In order to increase precision, it is necessary to measure the diameter of the pin during the turning process and the position of the cutter relative to the pin axis. It is desirable to fix the end point of the cutter underneath the axis of the pin nearly 10% of the diameter of the processed pin. Therefore, two tasks of measurement appear: the measurement of the pin diameter and the measurement of the cutter position relative to the pin axis. This work was accomplished by Alberto Caballero [54].

For the measurement of the pin diameter, the metallic disc that was located on the metallic bar (Fig. 8.36a) fixed on the lathe carriage was used. The metallic disc is placed underneath the measured pin. It rises (along the Z-axis) until electrical contact with the pin occurs. Thereafter, the measurement procedure is organized as follows:

1. The disc moves back along the X-axis and rises along the Z-axis with the step dZ (Fig. 8.36a).
2. The disc moves forward, and during this movement two conditions are verified: condition A – the disc makes electrical contact with the pin; condition B – the disc moves far away from the axis of the pin without electrical contact (Fig. 8.36b).
3. If condition A occurs, the coordinates (X, Z) are registered in the table of measurements, and the process returns to point 1.
4. If condition B occurs, the measurement process is finished.

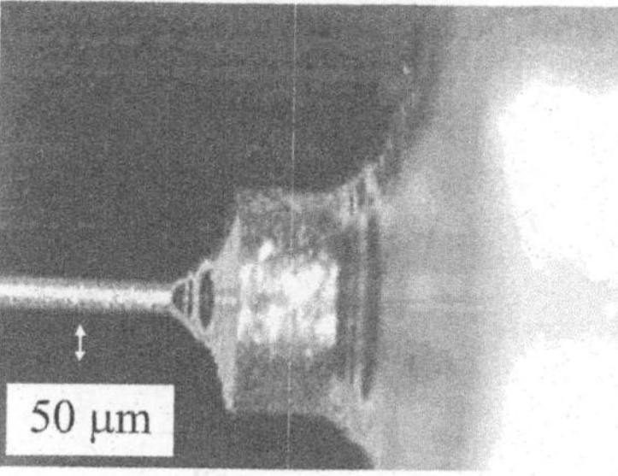

Fig. 8.35 Example of a brass pin of 50 μm diameter

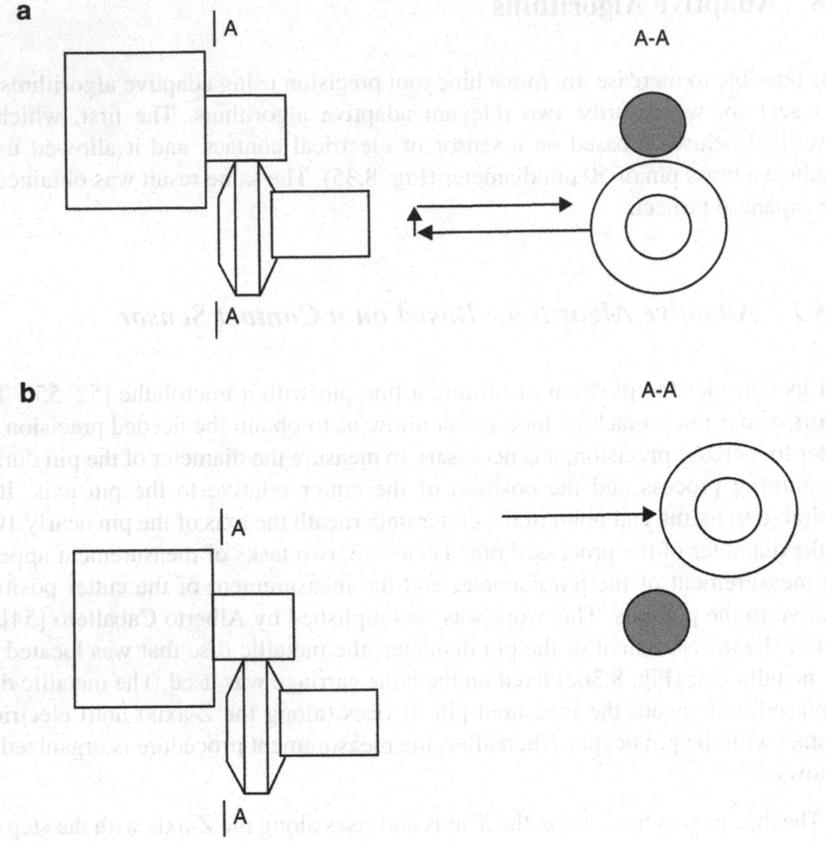

Fig. 8.36 Measurement of the diameter of the pin

The table of the measurement contains the coordinates of the points, which must lie on a circle (Fig. 8.37). In reality, as a result of measurement errors, these points can reside outside of the ideal circle (Fig. 8.38).

It is necessary to approximate the experimental points with the circle. The diameter of this circle will equal the sum of the pin diameter and unknown disc diameter. To construct the approximation circle, we used the following algorithm. Let X_0, Z_0 be the arbitrarily selected point that we considered as the approximation circle center. Let us find the medium distance r_i between the point X_0, Z_0 and experimental points.

$$r_i = \sqrt{(X_i - X_0)^2 + (Z_i - Z_0)^2}, \tag{8.111}$$

$$r_m = \frac{\sum_{i=1}^{k} r_i}{k}, \tag{8.112}$$

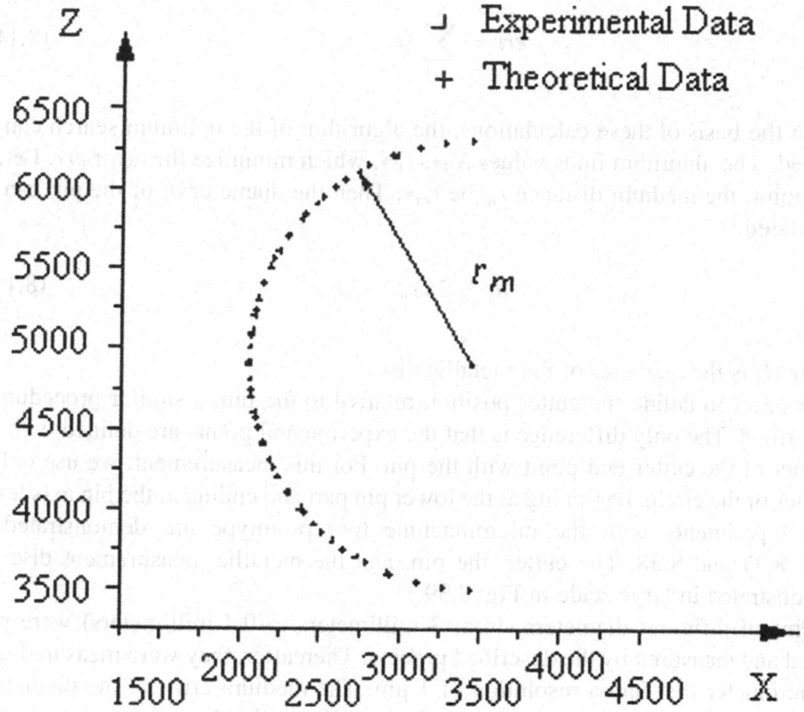

Fig. 8.37 Coordinates of the experimental points

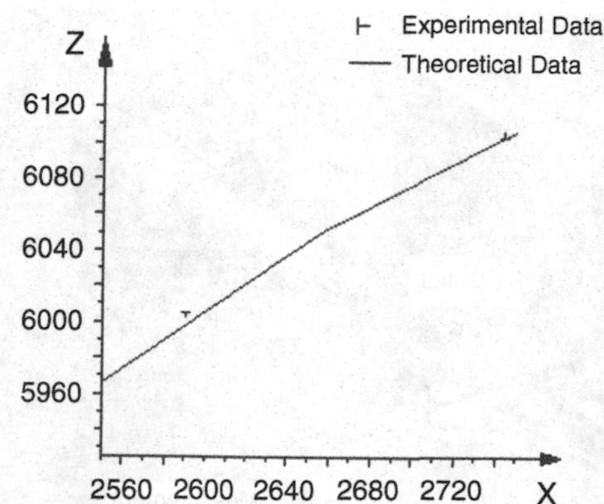

Fig. 8.38 Measurement errors

where k is the number of experimental points; X_i, Z_i are coordinates of the experimental points; r_i is the distance between point X_i, Z_i and point X_O, Z_O; and r_m is the medium distance. Further, it is necessary to calculate the quadratic error:

$$err = \sum_{i=1}^{k} (r_i - r_m)^2. \tag{8.113}$$

On the basis of these calculations, the algorithm of the optimum search can be created. The algorithm finds values X_0*, Y_0*, which minimize the error *err*. Let, in this point, the medium distance r_m be r_m*. Then the diameter d_s of the pin can be calculated:

$$d_s = 2 \cdot r_m^* - d_d, \tag{8.114}$$

where d_d is the diameter of the metallic disc.

In order to define the cutter position relative to the pin, a similar procedure is performed. The only difference is that the experimental points are defined with the contact of the cutter end point with the pin. For this measurement, we use only a quarter of the circle, beginning at the lower pin part and ending at the pin axis level. The experiments with the micromachine tool prototype are demonstrated in Figs. 8.37 and 8.38. The cutter, the pin, and the metallic measurement disc are demonstrated in large scale in Fig. 8.39.

Pins of different diameters (from 3 millimeters to 0.4 millimeters) were produced and measured by the described process. Thereafter, they were measured with a micrometer that had a resolution of 1 μm. The medium error of the diameter's value was 8 μm, and the maximum error was 22 μm. Without using the adaptive algorithm, the manufacturing errors reach 60 μm.

Fig. 8.39 Pin, cutter, and metallic measurement disc

The cutter positions relative to the pin axis were also measured. It is difficult to verify these positions directly, so in order to evaluate correctness, we produced extremely fine pins. If the cutter does not have the correct position, it is impossible to obtain a small diameter pin. We have made pins with a diameter as small as 50 μm. An example of such a pin is given in Fig. 8.35

8.9 Possible Applications of Micromachine Tools

Many applications of micromachine tools can be considered as a consequence of low cost, small space, and low energy consumption of such tools, and of the mass parallel automatic production process. The preferable area of application is the production of devices that contain huge numbers of micromechanical details of arbitrary 3D shapes. We suppose that many types of new products will appear due to the low cost of such devices. As examples, we can mention large flat screens with mechanically driven image cells, mechanically driven random access memory, and microbar assembly materials. Here we consider one such application: microfilters for fine filtration of liquids and gases.

This problem relates to different branches of technological processes and environmental protection. Fine filtration of air is needed for “clean rooms” used in microelectronics, medicine, and other applications. Filtration of gases is needed for environmental protection from pollution by power plants, transport vehicles, and other devices that use fossil fuel. Fine filters have many applications in the food and chemical industries, microbiology, and so on.

8.9.1 The Problem of Liquid and Gas Fine Filtration

Existing filters do not always solve filtration problems efficiently. Some of them do not perform properly; others do not give sufficiently fine filtration or cannot be used in all environmental conditions. Paper filters are often used for filtration of liquids and gases, and their main advantages are low cost and relatively high relation of throughput to pressure drops. At present, there are technologies for manufacturing filter papers with parallel fibres. Such filters permit the filtration of particles down to 0.2 μm, but they have some disadvantages, including the following:

- paper filters demand relatively large space;
- the paper cannot be used in some environmental conditions (e.g., high-temperature gases, liquids);
- paper filters for liquids do not permit filtration of particles smaller than 0.2 μm.

To filter particles smaller than 0.2 μm, membrane filters are used, but such filters have very low throughput and high pressure drops [57].

We propose to perform the fine filtration of liquids and gases having a high relation of throughput to pressure drop with a huge number of special microrings pressed on to each other. Filters based on such microrings can be made by MET. Below, we present the design of such filters.

8.9.2 Design of Filters with a High Relation of Throughput to Pressure Drop

The proposed filter contains a case, an inlet, an outlet, a flush valve, and a huge number of microrings assembled into columns (Fig. 8.40). Liquid (or gas) goes through the inlet and comes into the internal space of the microring columns. After this, it travels through slits between microrings and comes out through the outlet. Hard particles are held back in the internal space of the microring columns. To remove hard particles from the internal space of the microrings, it is necessary to open the flush valve. A liquid (or gas) flow directed from the inlet to the flush valve will remove the sediment. Ultrasound vibration could be applied for better cleaning of the internal space.

To estimate the main parameters of the filter, let us consider the slit between a pair of microrings (Fig. 8.41). Let the narrowest part of the slit have width W and the widths of other parts of the slit be calculated as:

$$x(r) = W + k_1 * (r_3 - r) \quad \text{for } r_3 > r \geq r_4, \tag{8.115}$$

$$x(r) = W + k_2 * (r - r_3) \quad for\ r_2 > r \geq r_3, \tag{8.116}$$

$$x(r) = W + k_2 * (r_2 - r_3) + k_3 * (r - r_2) \quad for\ r_1 > r \geq r_2, \tag{8.117}$$

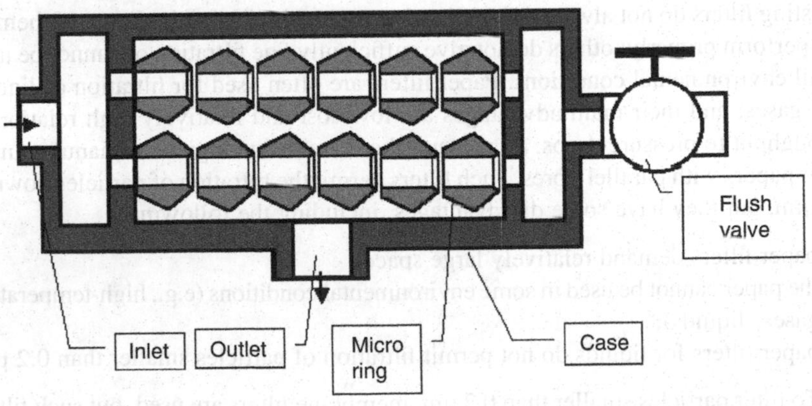

Fig. 8.40 Filter design

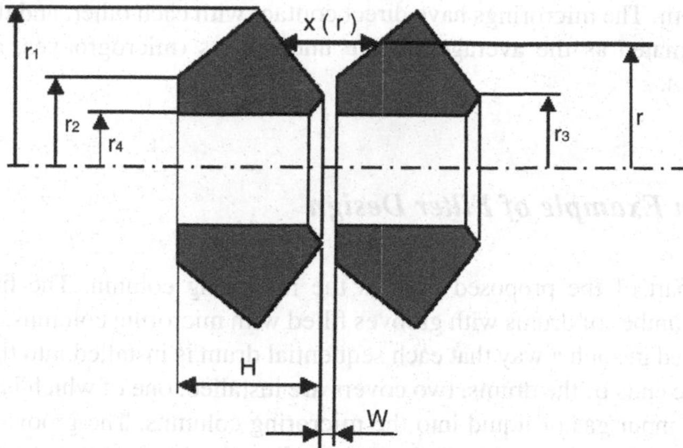

Fig. 8.41 Arrangement of rings

The main parameters of the filters are the following: the size of arrested particles, the relation of throughput to pressure drop, and the size (or the weight) of the filter. We have used a numerical model of the filtration process to calculate these properties.

To calculate the filter parameters, it is necessary to know the hydraulic resistance of one slit between two microrings:

$$G = P/Q, \tag{8.118}$$

where G is the hydraulic resistance of one slit between two microrings, P is the pressure drop in the slit, and Q is the flow rate in the slit.

The hydraulic resistance of the section between radius r and $r + dr$ can be calculated as

$$dG = \frac{6\mu}{\pi \cdot r \cdot x^3(r)} dr, \tag{8.119}$$

where dG is the hydraulic resistance, μ is the viscosity coefficient, r is the current radius (Fig. 8.41), and $x(r)$ is the current width of the slit at the radius r. $x(r)$ is calculated by expressions (8.115) – (8.117), and then

$$G = \int_{r_4}^{r_1} dG. \tag{8.120}$$

We have determined G by numerical integration in equation (8.120). Hydraulic resistance of the slit depends on the shape of the microrings and on the distance between them. The microrings have direct contact with each other, and the value W can be estimated as the average surface unevenness (microgrooves, absence of flatness, etc.).

8.9.3 An Example of Filter Design

The main part of the proposed filter is the microring column. The filter design contains a number of drums with grooves filled with microring columns. The drums are assembled in such a way that each sequential drum is installed into the previous drum. At the ends of the drums, two covers are installed, one of which has holes for sending the input gas or liquid into the microring columns. The grooves of neighboring drums are connected by holes for sending the output filtered gas or liquid into the internal space of the smallest drum connected to the output filter hole. Such a design may be used as a separate device or as part of a larger filter. To produce such a filter, it is necessary to manufacture two covers, a set of drums, and a huge number of microrings (Fig. 8.42).

8.9.4 The Problems of Fine Filter Manufacturing

Let us consider a filter having ten drums, with each drum containing 100 grooves and each groove containing 50 microrings. In this case, the total number of microrings equals 50,000. The package, covers, and drums have a very simple design that is not difficult to manufacture. The main problem of filter production, because of the huge number of microrings, is to manufacture microrings and install them into the drum grooves. We propose to manufacture the microrings by the usual methods of material cutting using computer numerical controlled cutting micromachine tools and to install the microrings into drum grooves using a computer numerical controlled assembly micromanipulator. It is practically impossible to organize such a process using traditional mechanical equipment. For this, it is necessary to develop micromechanical technology to produce the components of the microfilter [1].

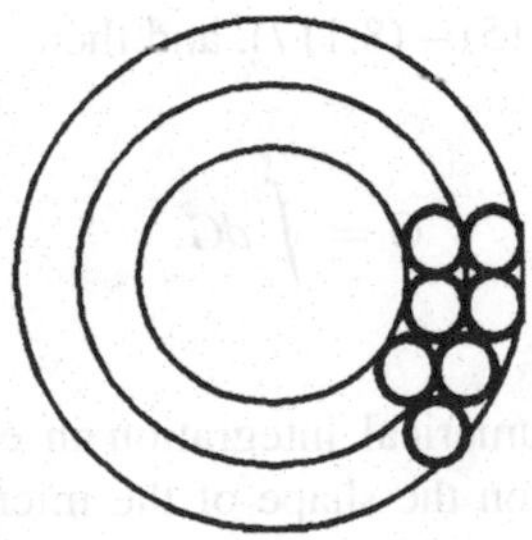

Fig. 8.42 Package of filter

8.9.5 The Filter Prototype Manufactured by the Second Micromachine Tool Prototype

To test the second micromachine tool prototype and examine the microfilter characteristics, we have made a microfilter prototype using the second micromachine tool prototype. A microfilter has one drum containing 12 columns with ten microrings in each column. The internal diameter of each microring is 0.8 mm, and the external diameter is 1.3 mm. The height of each microring is approximately 0.6 mm. (Fig. 8.43).

8.9.6 Case Study

We performed an experiment to study the microfilter's behavior. It consisted of measuring the flow rate at three different pressures using pure water. The results obtained were compared with theoretical results, calculated by software. The final results are shown in Table 8.7 and Fig. 8.44.

We have begun investigating a new approach to micromechanical device fabrication, proposed in [1]. The approach assumes the production of microequipment (micromachine tools and micromanipulators) as sequential generations. Overall sizes of future generations of microequipment must be smaller than the overall sizes of the previous generations. To date, we have made two prototypes of micromachine tools of the first generation.

We have worked out some other elements needed for microfactory creation. There is a micromanipulator for loading and unloading micromachine tools, a

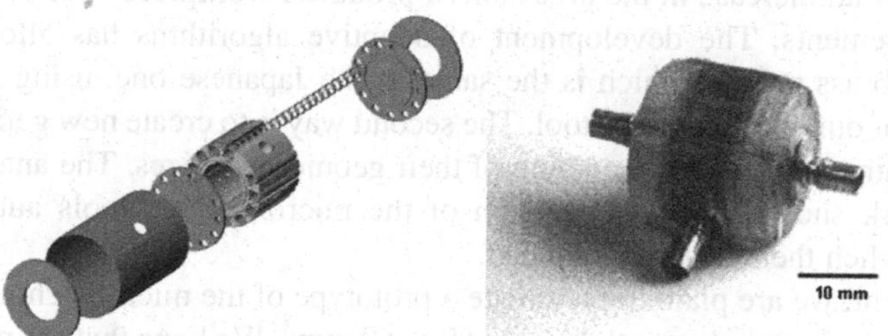

Fig. 8.43 Filter assembly and prototype

Table 8.7 Results of flow rate with different pressures

	Pressure [Kpa]	Real Flow Rate [mm^3/s]	Theoretical Flow Rate [mm^3/s]
Experiment 1	3.237	2.33	2.9
Experiment 2	5.101	3.68	4.5
Experiment 3	7.063	5.09	5.6

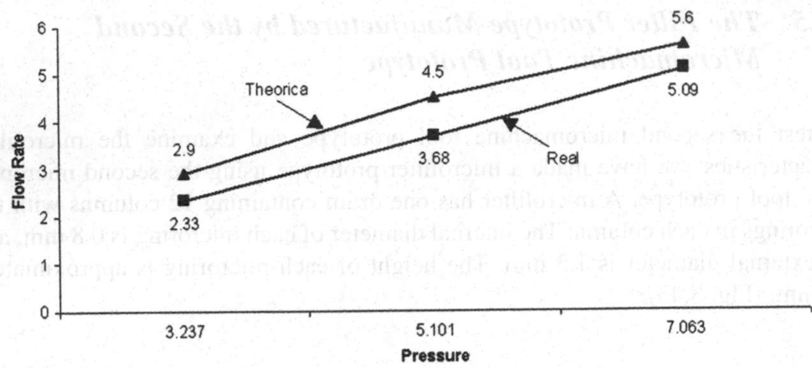

Fig. 8.44 Flow rate versus pressure in the microfilter

micromachine tool for microassembly under control of a technical vision system, and a parallel system of control for many micromachine tools and micromanipulators with the use of one PC [58]. All these developments are based on inexpensive components with the possibility of future size reduction. Constructed prototypes have shown that it is possible to create micromachine tools with components costing less than $100. These micromachine tools differ from others by the simplicity of their design and production. The technology of micromachine tool manufacture provides the opportunity of size reduction in future generations.

A comparison of the developed prototypes of micromachine tools with similar Japanese micromachine tools shows that the component cost of our prototypes is much less than that of the Japanese equipment. However, the developed prototypes do not have such high precision as the Japanese ones. There are two ways to compensate for this imperfection. One way is to develop adaptive algorithms, which allow an increase in the precision of produced workpieces due to intermediate measurements. The development of adaptive algorithms has allowed us to produce a brass needle, which is the same as the Japanese one, using the second prototype of our micromachine tool. The second way is to create new generations of micromachine tools with a reduction of their geometrical sizes. The analysis made in this work shows that the precision of the micromachine tools automatically increases when their sizes are reduced.

At present, we are planning to create a prototype of the micromachine tool with overall dimensions no larger than $60 \times 60 \times 60\ mm^3$. We hope that the precision of this micromachine tool will be two or three times higher. In the future, we will continue to utilize manufacturing technology to reduce the micromachine tool sizes and increase their precision.

MET can be applied to almost all small-sized mechanical devices. The main principle of application is that the sizes of machine tools must be comparable with the sizes of the produced microcomponents. For example, the modern electronic industry produces large amounts of connectors, each containing many complementary "hole-pin" pairs. The sizes of these parts are usually equal to 1–5 mm. All these parts could be produced with machine tools having sizes of 10–50 mm. The low-cost technologies of stamping, forging, and so on can be fully reproduced in equipment of this size.

Another example of MET application in microelectronics is rapid prototyping of multi-chip modules (MCMs). For control systems of our microequipment, we have developed and checked new rapid prototyping technology, which uses magnet wire connections for wire board manufacturing [59]. Miniaturization of the equipment should permit the application of this technology to multi-chip module prototyping. If a multi-chip module size is 20×20 mm^2, the machine tool sizes could be $200 \times 200 \times 200$ mm^3 (the first-generation machine tools).

At present, many devices (photo camera, TV camera, etc.) use DC motors with sizes as low as 10 mm. All the machine tools and assembly devices for producing such motors could have sizes less than $100 \times 100 \times 100$ mm^3, preserving the conventional technology of DC motor manufacture.

8.10 Conclusion

We propose the use of all types of conventional mechanical technologies for producing micromechanical devices. For this purpose, it is necessary to reduce the overall size of machine tools and assembly devices proportionally to the size of the devices to be produced. These devices should be made by sequential generations of micromechanical equipment of smaller and smaller sizes. Each new generation should be produced using previous generations. The experience in producing and evaluating prototypes of first-generation micromachine tools, described here, shows that our goal is realistic. Conventional mechanical technologies can be transferred to the microworld by miniaturization of the corresponding technological equipment. Some prototypes of the first-generation equipment have been developed. These prototypes have low-cost components (less than $100), simple design, and the possibility of scaling down. Test pieces manufactured with these prototypes have dimensions from 50 μm to 2 mm. The tolerance of the detail dimensions is in the neigborhood of 20 μm. Additional efforts are needed to decrease the tolerances while preserving the low cost of the machine tool.

The low cost of the proposed micromachine tools is one of the main advantages of microequipment technology. Other advantages include the 3D shape of produced details, the use of any materials, and the possibility of using, in the microdomain, the designs and manufacturing principles developed for macrodevices.

A theoretical analysis of the machine tool errors made in this chapter shows that it is possible to make microequipment as a series of sequential generations. Each subsequent generation is smaller than the previous generation and could be produced with the microequipment of previous generations. The majority of microequipment errors decrease proportionally with reductions in microequipment size. However, there are exceptional errors that do not decrease automatically. To minimize such errors, special technological processes and adaptive algorithms of manufacturing are proposed. These additional methods allow us to scale down exceptional errors, too. Such properties make it possible to scale down micromachine tools and micromanipulators without design changes and without the use of special super-precise methods to obtain the required tolerances. This approach should allow us to create low-cost and effective microequipment for micromechanical device manufacturing.

References

1. Kussul E.M., Rachkovskij D.A., Baidyk T.N., et al., Micromechanical Engineering: a Basis for the Low Cost Manufacturing of Mechanical Microdevices Using Microequipment. *J. Micromech. Microeng*, 1996, Vol. 6, pp. 410–425.
2. Ohlckers P., Hanneborg A., Nese M., Batch Processing for Micromachined Devices. *J. Micromech. Microeng*, 1995, Vol. 5, pp. 47–56.
3. Handbook of Microlithography, Micromachining, and Microfabrication. Vol. 2: Micromachining and Microfabrication. Ed. by P. Rai-Choundhury. SPIE Press, 1997.
4. Micromechanics and MEMS. Classsical and Seminal Papers to 1990. Ed. by W. S. Trimmer, IEEE Press, New York, 1997.
5. Kussul E., Baidyk T., Rachkovskij D., Talaev S. The Method of Micromechanical Manufacture. Patent No. 2105652, Russia, 27.02.1998, priority from 2.02.1996 (in Russian).
6. Kussul E., Baidyk T., Rachkovskij D., Talaev S. The Method of Micromechanical Manufacture. Patent No. 24091, Ukraine, 31.08.1998, priority from 17.01.1996 (in Ukrainian).
7. Naotake Ooyama, Shigeru Kokaji, Makoto Tanaka et al., Desktop Machining Microfactory. *Proceedings of the 2-nd International Workshop on Microfactories, Switzerland*, Oct.9–10, 2000. pp. 13–16.
8 Kussul E., Baidyk T., Ruiz-Huerta L., Caballero A., Velasco G., Kasatkina L. Development of Micromachine Tool Prototypes for Microfactories, *Journal of Micromechanics and Microengineering*, 12, 2002, pp. 795–812.
9. Ohlckers P., Jakobsen H. High Volume Production of Silicon Sensor Microsystems for Automotive Applications. *IEEE Colloquium on Assembly and Connection in Microsystems* (Digest No. 1997/004), 1997, pp. 8/1–8/7.
10. Eddy D.S., Sparks D.R. Application of MEMS Technology in Automotive Sensors and Actuators. *Proceedings of the IEEE*, Vol.86, Issue 8, 1998, pp. 1747–1755.
11. Tang T.K., Gutierrez R.C., Stell C.B., Vorperian V., Arakaki G.A., Rice G.T., Li W.J., Chakraborty I., Shcheglow K., Wilcox J.Z., Kaiser W.J. A Packaged Silicon MEMS Vibratory Gyroscope for Microspacecraft. *10th Annual International Workshop on Micro Electro Mechanical Systems*, 1997, pp. 500–505.
12. Norvell B.R., Hancock R.J., Smith J.K., Pugh M.L., Theis S.W., Kriatkofsky J. Micro Electro Mechanical Switch (MEMS) Technology Applied to Electronically Scanned Arrays for Space Based Radar. *Aerospace Conference*, 1999, Proceedings, Vol. 3, pp. 239–247.
13. Madni A.M., Wan L.A. Micro Electro Mechanical Systems (MEMS): an Overview of Current State-of-the Art. *Aerospace Conference*, 1998 IEEE, Vol. 1, pp. 421–427.
14. Janson S., Helvajian H., Amimoto S., Smit G., Mayer D., Feuerstein S. Microtechnology for Space Systems. *Aerospace Conference*, 1998 IEEE, Vol. 1, pp. 409–418.
15. Minor R.R., Rowe D.W. Utilization of GPS/MEMS-IMU for Measurement of Dynamics for Range Testing of Missiles and Rockets. *Position Location and Navigation Symposium*, IEEE 1998, pp. 602–607.
16. Yu-Chong Tai. Aerodynamic Control of a Delta-wing Using MEMS Sensors and Actuators. *Proceedings of the 1997 International Symposium on Micromechatronics and Human Science*, 1997, pp. 21–26.

17. Fujita H. Microactuators and Micromachines. *Proceedings of the IEEE*, Vol. 86, 1998, No. 8, pp. 1721–1732.
18. Comtois J.H., Michalicek M.A., Clark N., Cowan W., MOEMS for Adaptive Optics. Broadband Optical Networks and Technologies: An Emerging Reality/Optical MEMS/ Smart Pixels/Organic Optics and Optoelectronics, 1998, IEEE/LEOS Summer Topical Meetings, 1998, pp. II/95–II/96.
19. McCormick F. B, Optical MEMS Potentials in Optical Storage. Broadband Optical Networks and Technologies: An Emerging Reality/Optical MEMS/Smart Pixels/Organic Optics and Optoelectronics, 1998, IEEE/LEOS Summer Topical Meetings, pp. II/5–II/6.
20. Kuwano H., MEMS for Telecommunication Systems. *Proceedings of the 7th International Symposium on Micro Machine and Human Science*, 1996, pp. 21–28.
21. Johnson M. D. Hughes G. A. Gitlin M. L. Loebel N. G. Paradise N. F. Cathode Ray Addressed Micromirror Display. *Proceedings of the 13th Biennial University/ Government/ Industry Microelectronics Symposium*, 1999, pp. 158–160.
22. Brown E.R., RF-MEMS Switches for Reconfigurable Integrated Circuits, *IEEE Transactions on Microwave Theory and Techniques*, Vol. 46, No. 11, 1998, pp. 1868–1880.
23. Pillans B., Eshelman S., Malczewski A., Ehmke J., Goldsmith C., Ka-band RF MEMS Phase Shifters. *IEEE Microwave and Guided Wave Letters*, 1999, Vol. 9, pp. 520–522.
24. Wu H.D., Harsh K.F. Irwin R.S. Wenge Zhang, Mickelson A.R., Lee Y.C., Dobsa J.B., MEMS Designed for Tunable Capacitors. *Microwave Symposium Digest*, 1998 IEEE MTT-S International, 1998, Vol. 1, pp. 127–129.
25. Wu S., Mai J., Tai Y.C., Ho C.M., Micro Heat Exchanger by Using MEMS Impinging Jets. *12th IEEE International Conference on Micro Electro Mechanical Systems*, 1999, pp. 171–176.
26. Rachkovskij D. A., Kussul E. M., Talayev S. A., Heat Exchange in Short Microtubes and Micro Heat Exchangers with Low Hydraulic Losses. *Microsystem Technologies*, 1998, Vol. 4, pp. 151–158.
27. Bier W., Keller W., Linder G., Siedel D., Schubert K., Martin H. Gas-to Gas Transfer in Micro Heat Exchangers. *Chemical Engineering and Processing*, 1993, Vol. 32, pp. 33–43.
28. Friedrich C., Kang S. Micro Heat Exchangers Fabricated by Diamand Machining. *Precision Engineering*, 1994, Vol. 16, pp. 56–59.
29. Katsura S., Hirano K., Yamaguchi A., Ishii R., Imayou H., Matsusawa Y., Mizuno A. Manipulation of Chromosomal DNA and Localization of Enzymatic Activity. *Industry Applications Conference*, 32nd IAS Annual Meeting IAS'97, Conference Record of the 1997 IEEE, 1997, Vol. 3, pp. 1983–1989.
30. Dohi T. Computer Aided Surgery and Micro Machine. *Proceedings of the 6th International Symposium on Micro Machine and Human Science*, MHS'95, 1995, pp. 21–24.
31. Cibuzar G., Polla D., McGlennen R., Biomedical MEMS Research at the University of Minnesota. *Proceedings of the 12th Biennial University/Government/Industry Microelectronics Symposium*, 1997, pp. 145–149.
32. Sun L., Sun P., Qin X., Wang C., Micro Robot in Small Pipe with Electromagnetic Actuator. *International Symposium on Micromechatronics and Human Science*, 1998, pp. 243–248.
33. Caprari G., Balmer P., Piguet R., Siegwart R., The Autonomous Micro Robot "Alice": a Platform for Scientific and Commercial Applications. International Symposium on Micromechatronics and Human Science, 1998, pp. 231–235.
34. Hayashi I., Iwatsuki N., Micro Moving Robotics. *International Symposium on Micromechatronics and Human Science*, 1998, pp. 41–50.
35. http://www.siliconstrategies.com/story/OEG20020827S0031
36. Micro Mechanical Systems: Principles and Technology. Ed. By T. Fukuda and W. Menz. Elsevier Science B.V. 1998.
37. Mazuzawa T., An Approach to Micromachining through Machine Tool Technology. *Proc. 2nd Int. Symp. Micro Machine and Human Science* (Nagoya, Japan), 1991, pp. 47–52.
38. Friedrich C. R. and Vasile M. J., Development of the Micromilling Process for High- Aspect-Ratio Micro Structures. *J. Microelectromechanical Systems*, 1996, 5, pp. 33–38.

39. Friedrich C.R. and Kang S.D., Micro Heat Exchangers Fabricated by Diamond Machining. *Precision Engineering*, 1994, 16, pp. 56–59.
40. Yamagata Y. and Higuchi T. Four Axis Ultra Precision Machine Tool and Fabrication of Micro Parts by Precision Cutting Technique. *Proc. 8th Int. Precision Engineering Seminar* (Compiegne, France), 1995, pp. 467–470.
41. Ishihara H., Arai F., Fukuda T., Micro Mechatronics and Micro Actuators. *IEEE/ASME Transactions on Mechatronics*, 1996, Vol. 1, pp. 68–79.
42. Some Micro Machine Activities in Japan. Report ATIP96.021, 1996.
43. Okazaki Yuichi, Kitahara Tokio, Micro-Lathe Equipped with Closed-Loop Numerical Control, *Proceedings of the 2nd International Workshop on Microfactories*, Switzerland, Oct. 9–10, 2000, pp. 87–90.
44. Maekawa H., Komoriya K. Development of a Micro Transfer Arm for a Microfactory. *Proceedings of the 2001 IEEE International Conference on Robotics & Automation*, Seoul, Korea, May 2001, pp. 1444–1451.
45. Bleuler H., Clavel R., Breguet J.-M., Langen H., Pernette E. Issues in Precision Motion Control and Microhandling. *Proceedings of the 2000 IEEE International Conference on Robotics & Automation*, San Francisco, 2000, pp. 959–964.
46. Ishikawa Yu., Kitahara T. Present and Future of Micromechatronics. *1997 International Symposium on Micromechatronics and Human Science*, 1997, pp.13–20.
47. Naotake Ooyama, Shigeru Kokaji, Makoto Tanaka et al., Desktop Machining Microfactory. *Proceedings of the 2-nd International Workshop on Microfactories*, Switzerland, Oct. 9–10, 2000, pp. 14–17.
48. Trimmer W.S.N. Microrobots and Micromechanical Systems. In: Sensors and Actuators, 19, 1989, pp. 267–287.
49. Kussul E., Baidyk T., Ruiz-Huerta L., Caballero-Ruiz A., Velasco G., Makeyev O., Techniques in the Development of Micromachine Tool Prototypes and Their Applications in Microfactories. In: MEMS/NEMS Handbook: Techniques and Applications, Ed. by Cornelius T. Leondes, Kluwer Academic Publishers, 2006, Vol. 3, Chapter 1, pp. 1–61.
50. Kussul E., Baidyk T., Ruiz-Huerta L., Caballero-Ruiz A., Velasco G., Scaling Down of Microequipment Parameters, *Precision Engineering*, 2006, Vol. 30, Issue 2, pp. 211–222.
51. G. H. J. Florussen, F. L. M. Delbressine, M. J. G. van de Molengraft, P. H. J. Schellekens, Assessing Geometrical Errors of Multi-Axis Machines by Three-Dimensional Length Measurement. *J. Measurement*, 30, 2001, pp. 241–255.
52. Kussul E., Baidyk T., Ruiz-Huerta L., Caballero-Ruiz A., Velasco G., CNC Micromachine Tool: Design & Metrology Problems, in: Advances in Systems Theory, Mathematical Methods and Applications, A. Zemliak, N. Mastorakis, (eds.), Greece, 2002, WSEAS Press, pp. 93–97.
53. Caballero-Ruiz A., Ruiz-Huerta L., Kussul E., Baidyk T., Velasco G., Micromachine Tool: Measurement and Control, 17th Annual Meeting, ASPE, St. Louis, Missouri, Oct. 20–25, 2002, pp. 377–382.
54. Alberto Caballero Ruiz, Metodología de Evaluación para Microequipo, Doctorado en Ingeniería Mecánica (Tesis), 11 May 2005, pp. 116.
55. Caballero-Ruiz A., Ruiz-Huerta L., Baidyk T., Kussul E. Geometrical Errors Analysis of a CNC Micro-Machine Tool, *Mechatronics*, Vol. 17, Issues 4–5, May–June 2007, pp. 231–243.
56. Leopoldo Ruiz Huerta, Desarrollo de Microequipo para Células de Manufactura, Doctorado en Ingeniería Mecánica (Tesis), 11 May 2005, pp. 110.
57. Yang X., Yang M.J., Wang X.O., Meng E., Tai Y.C., Ho C.M. Jan. 1998 Micromachined Membrane Particle Filters. *Proceedings of The Eleventh Annual International Workshop on Micro Electro Mechanical Systems, MEMS 98*, 1998, pp. 137–142.
58. Kussul E. Control Paralelo de Máquinas Herramientas, *Congreso SOMI'XVI*, Querétero, México, October 15–19, 2001, pp. 7.
59. Kussul E., López Walle B. Tecnología de Alambrado por Tejido Alambre Magneto. *Congreso SOMI XVII*, Mérida, Yucatán, México, 14–18 October, 2002, pp. 8.

Chapter 9
Applications of Neural Networks in Micromechanics

A computer vision system permits one to provide feedback, which increases the precision of the manufacturing process. It could be used in low-cost micromachine tools and micromanipulators for microdevice production. A method of sequential generations was proposed to create such microequipment [1], Chapter 8.

9.1 Neural-Network-Based Vision System for Microworkpiece Manufacturing

According to this method, the microequipment of each new generation is smaller than the equipment of the previous generations. This approach would allow us to use low-cost components for each microequipment generation and to create microfactories capable of producing low-cost microdevices [2]. To preserve high precision of the microequipment, it is necessary to use adaptive algorithms of micropiece production. Algorithms based on contact sensors were proved and showed good results [2]. The neural-network-based vision system could provide much more extensive possibilities to improve the manufacturing process.

As described in Chapter 8, a special project for microfactory creation based on miniature micromachine tools was initiated in Japan [3]. The Mechanical Engineering Laboratory developed a desktop machining microfactory [4, 5] consisting of machine tools such as a lathe, a milling machine, a press machine, and assembly machines such as a transfer arm and a two-fingered hand. This portable microfactory has external dimensions of 625 x 490 x 380 mm^3.

The idea of microfactory creation is also supported in other countries. In Switzerland, the details of precision motion control and microhandling principles for future microfactories have been worked out [6]. One of the main problems in such microfactories is that of automation on the basis of vision systems. There are different approaches to constructing computer vision systems [7–9] for these purposes.

E. Kussul et al., *Neural Networks and Micromechanics*
DOI 10.1007/978-3-642-02535-8_9, © Springer-Verlag Berlin Heidelberg 2010

In this section, we communicate the preliminary results of neural-network-based vision system investigation. We have proved the system in off-line mode. For this purpose, we have manufactured four groups of screws with different positions of the machine tool cutter. The images of these screws were used to train and test the developed system.

The task of shape recognition is well known [10]. The recognition of screw images, in our case, is based on the recognition of the screw shape or profile. We detected the screw contours and used this presentation as input for the recognition system. The vision system was based on the neural network with the permutation coding technique described in [11, 12]. This type of neural network showed good results in handwriting and face recognition tasks. Now we prove it for adaptive algorithms in manufacturing [13].

9.2 The Problems of Adaptive Cutting Processes

To increase the precision of micromachine tools, it is possible to use adaptive cutting processes [2]. Let us consider a lathe equipped with one TV camera, which could be moved automatically from position 1 to position 2 (Fig. 9.1). The images from the TV camera in the first position could be used to measure the sizes of partially-treated workpieces and to make the necessary corrections in the cutting process. The images from the TV camera in the second position could be used to correct the position of the cutting tool relative to the workpieces (Fig. 9.2).

In both positions, the TV camera can give useful information about the passing of the cutting process, for example, about chip formation, the contact of the cutter with the workpiece, and so on. All such images must be treated with the image recognition system. We propose to create such a recognition system on the basis of a neural network with permutative coding.

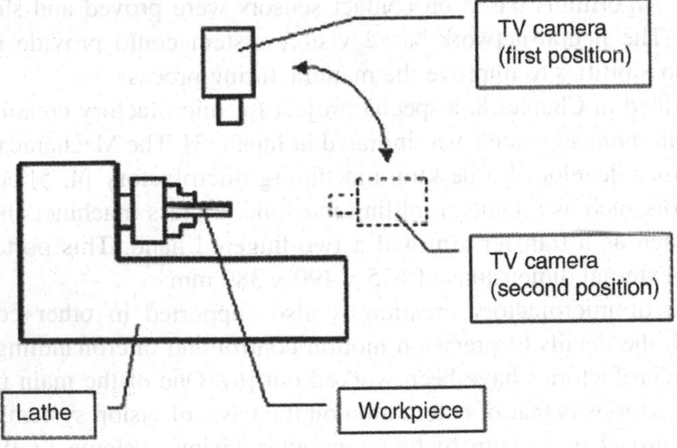

Fig. 9.1 The lathe equipped with TV camera

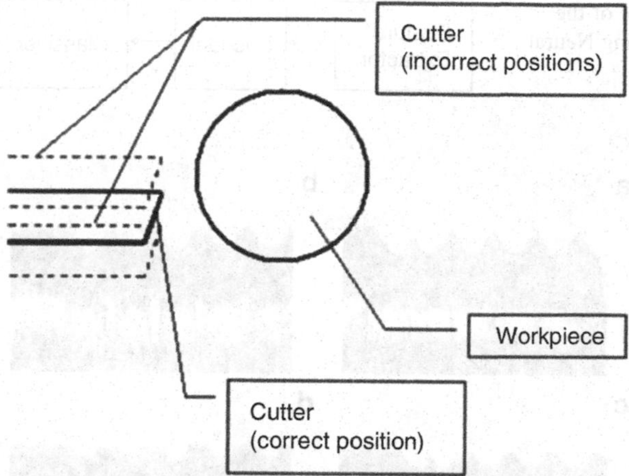

Fig. 9.2 The position of the cutting tool relative to the workpiece

9.3 Permutation Coding Neural Classifier

A Permutative Coding Neural Classifier (PCNC) was developed for the recognition of different types of images. It was proved on the MNIST database (handwritten digit recognition) and the ORL database (face recognition) and showed good results [11, 12]. Here, we examine this classifier in micromechanical applications [13]. The PCNC structure is presented in Fig. 9.3. The image is input to the feature extractor. The extracted features are applied to the encoder input. The encoder produces an output binary vector of large dimension, which is presented to the input of the one-layer neural classifier. The classifier output gives the recognized class.

9.3.1 Feature Extractor

An initial image (Fig. 9.4) is to be input to the feature extractor. The feature extractor begins with the selection of specific points in the image. In principle, various methods of selecting the specific points could be proposed. For example, the contour points could be selected as the specific points. The rectangle of $h \bullet w$ size is formed around each specific point (Fig. 9.5).

The specific points were selected using the following procedure. For each set of four neighboring pixels, we calculate the following expressions:

$$\begin{aligned} d_1 &= \left|br_{ij} - br_{i+1j+1}\right|, \\ d_2 &= \left|br_{ij+1} - br_{i+1j}\right|, \\ \Delta &= \max(d_1, d_2), \end{aligned} \tag{9.1}$$

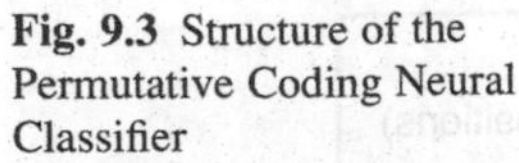

Fig. 9.3 Structure of the Permutative Coding Neural Classifier

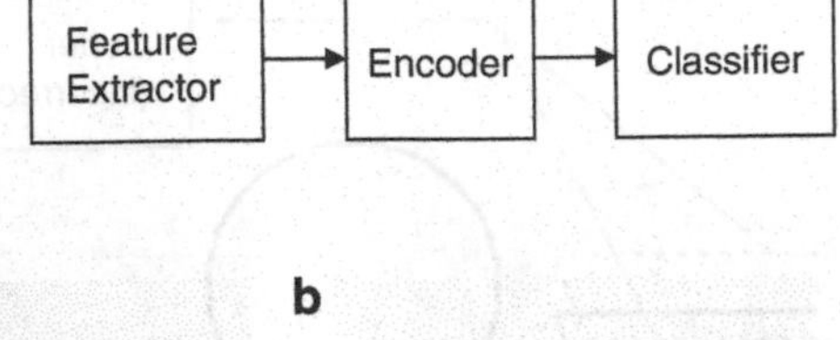

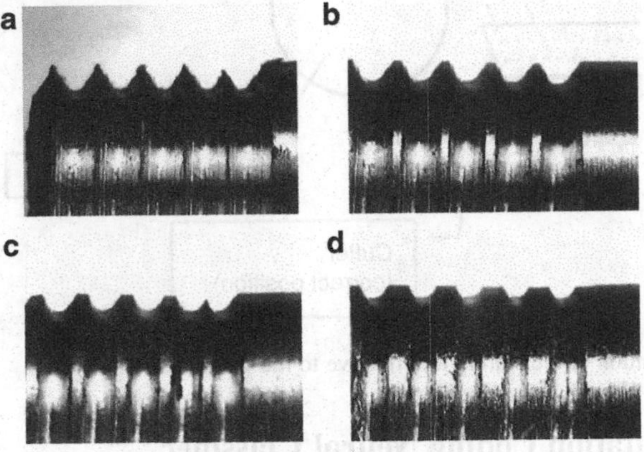

Fig. 9.4 Initial images

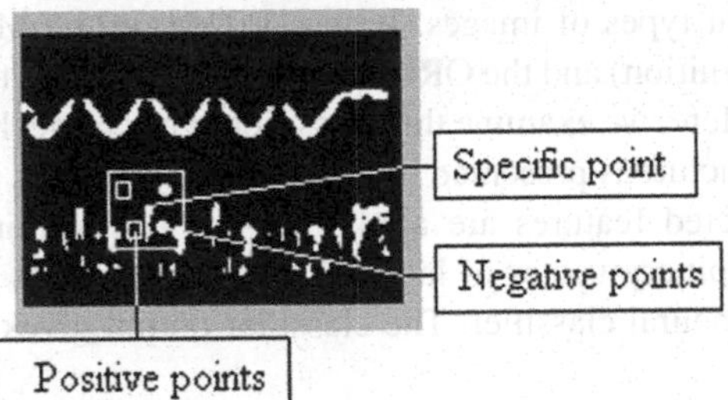

Fig. 9.5 Specific points selected by the feature extractor

where br_{ij} is the brightness of the pixel (i, j). If $(\Delta > B)$, then the pixel (i, j) corresponds to the selected specific point on the image, where B is the threshold for specific point selection.

Each feature is extracted from the rectangle of $h{\bullet}w$ size (Fig. 9.5). The p positive and the n negative points determine one feature and are randomly distributed in the rectangle. Each point P_{rs} has the threshold T_{rs} that is randomly selected from the range:

$$T_{min} \leq T_{rs} \leq T_{max} \tag{9.2}$$

where s stands for the feature number, and r stands for the point number.

The positive point is "active" if, in the initial image, it has brightness

$$b_{rs} \geq T_{rs}. \tag{9.3}$$

The negative point is "active" if it has brightness

$$b_{rs} \leq T_{rs}. \tag{9.4}$$

The feature under investigation exists in the rectangle if all its positive and negative points are active. In other cases, the feature is absent in the rectangle.

9.3.2 Encoder

The encoder transforms the extracted features to the binary vector:

$$V = \{v_i\} \ (i = 1, \ldots, N),$$

where $v_i = 0$ or 1. For each extracted feature F_s, the encoder creates an auxiliary binary vector:

$$U = \{u_i\} \ (i = 1, \ldots, N),$$

where $u_i = 0$ or 1.

A special random procedure is used to obtain the positions of the 1s in the vector U_s for each feature F_s. This procedure generates the list of the positions of "1" for each feature and saves all such lists in the memory. The vector U_s we term the "mask" of the feature F_s. To create this vector, it is necessary to take the positions from the list, fill them with "1," and fill other positions with "0."

In the next stage of the encoding process, it is necessary to transform the auxiliary vector U to the new vector U^* that corresponds to the feature location in the image. This transformation is made with permutations of the vector U components (Fig. 9.6). The number of permutations depends on the feature location in the image. The permutations in horizontal (X) (Fig. 9.6a) and vertical (Y) (Fig. 9.6b) directions are different permutations.

Each feature can have different locations in the image. The feature will have a different binary code for each location. For two locations of the same feature, the binary codes must be strongly correlated if the distance between the feature locations is small and must be weakly correlated if the distance is large. Such properties could be obtained with the following procedure. To code the location of feature F_s in the image, it is necessary to select the correlation distance D_c and calculate the following values:

$$\begin{aligned} X &= j \ / \ Dc, \\ E(X) &= (int)X, \\ R(X) &= j - E(X) \bullet Dc, \end{aligned} \tag{9.5}$$

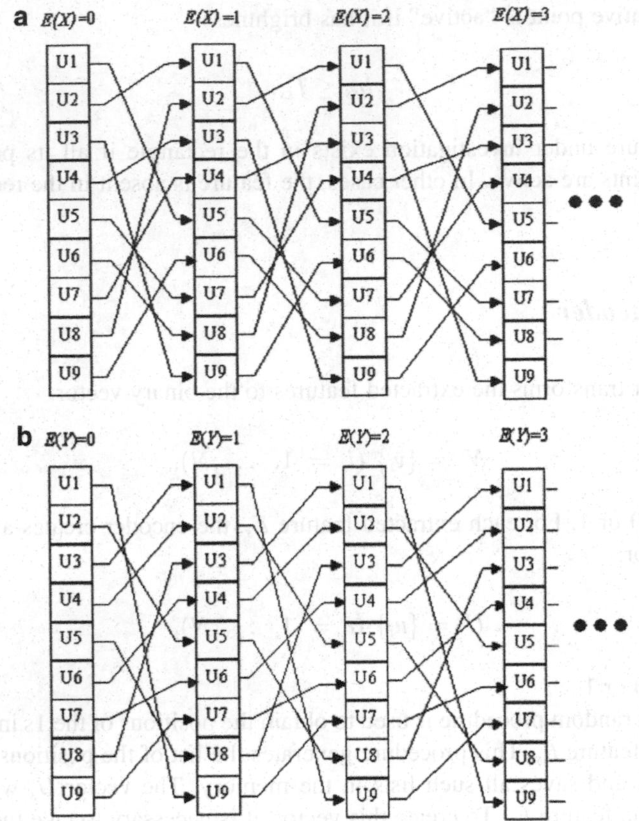

Fig. 9.6 Permutation pattern

$$\begin{aligned} Y &= i \,/\, D_c, \\ E(Y) &= (int)Y \\ R(Y) &= i - E(Y) \bullet D_c, \end{aligned} \tag{9.6}$$

$$P_x = \frac{R(X) \bullet N}{D_c}, \tag{9.7}$$

$$P_y = \frac{R(Y) \bullet N}{D_c}, \tag{9.8}$$

where $E(X)$ is the integer part of X; $R(X)$ is the fraction part of X; i is the vertical coordinate of the detected feature; j is the horizontal coordinate of the detected feature; and N is the number of neurons.

The mask of the feature F_s is considered as a code of this feature located at the top left corner of the image. To shift the feature location in the horizontal direction, it is necessary to make its permutations $E(X)$ times and to make an additional permutation for P_x components of the vector. After that, it is necessary to shift the code to the vertical direction, making the permutations $E(Y)$ times and an additional permutation for P_y components.

9.3.3 Neural Classifier

The structure of the proposed recognition system is presented in Fig. 9.7. The system contains sensor layer S, feature extractor, encoder, associative neural layer A, and reaction neural layer R. In the screw shape recognition task, each neuron of the R-layer corresponds to one of the screw groups from the database. The sensor layer S corresponds to the initial image.

The associative neural layer contains "binary" neurons having the outputs 0 or 1. The values of its neuron outputs are produced as a result of the encoder's work. The neurons of the associative layer A are connected to the reaction layer R with trainable connections having the weights w_{ji}. The excitations of the R-layer neurons are calculated in accordance with the equation:

$$E_i = \sum_{j=1}^{n} a_j * w_{ji}, \tag{9.9}$$

where E_i is the excitation of the ith neuron of the R-layer; a_j is the excitation of the jth neuron of the A-layer; and w_{ji} is the weight of the connection between the jth neuron of the A-layer and the ith neuron of the R-layer. The neuron-winner having maximal excitation is selected after calculating the excitations.

We use the following training procedure. Denote the neuron-winner number as i_w, and the number of the neuron, which really corresponds to the input image, as i_c. If $i_w = i_c$, nothing must be done. If $i_w \neq i_c$

$$\begin{gathered} (\forall j)\left(w_{ji_c}(t+1) = w_{ji_c}(t) + a_j\right) \\ (\forall j)\left(w_{ji_w}(t+1) = w_{ji_w}(t) - a_j\right) \\ \text{if}\left(w_{ji_w}(t+1) < 0\right) w_{ji_w}(t+1) = 0, \end{gathered} \tag{9.10}$$

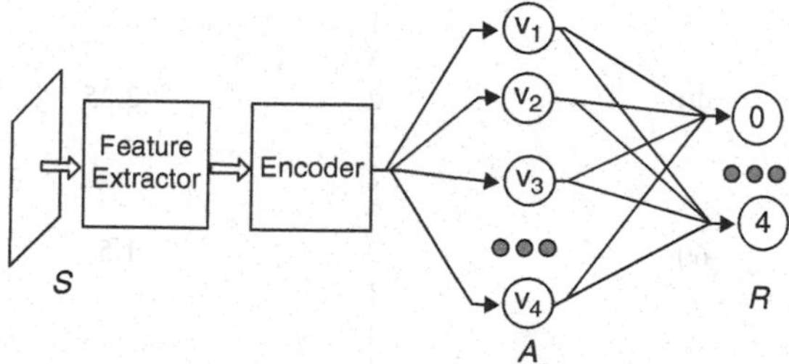

Fig. 9.7 Permutative Coding Neural Classifier

where $w_{ji}(t)$ is the weight of the connection between the j neuron of the A-layer and the i neuron of the R-layer before reinforcement, and $w_{ji}(t+1)$ is the weight after reinforcement.

9.4 Results

To examine the PCNC's recognition of shapes of micromechanical workpieces, we have produced 40 screws of 3 mm diameter with a CNC lathe from the "Boxford" company. Ten screws were produced with correct positioning of the thread-cutting cutter (Fig. 9.4b). 30 screws were produced with erroneous positioning of this cutter. Ten of them (Fig. 9.4a) had a distance of 0.1 mm less than necessary between the cutter and screw axis. Ten screws (Fig. 9.4c) were produced with a distance of 0.1 mm larger than necessary, and the remaining ten (Fig. 9.4d) with a distance of 0.2 mm larger than necessary. We have made the image database from these screws using a Samsung Web-camera mounted on an optical microscope.

Five images from each group of screws were selected randomly and used for neural classifier training, and another five were used for neural classifier examination. The experiments were conducted with different values of parameter B for specific point selection. In Table 9.1, the first column corresponds to parameter B. Four different runs were made for each value of B to obtain statistically reliable results. Each run differs from the others by the set of samples selected randomly for neural classifier training and by the permutation scheme structure. For this reason, we obtained different error rates in different runs. The second column contains the error number for each run. The third column gives the mean value of the error number for each B value. The fourth column contains the mean percentage of correct recognition.

A new neural classifier, PCNC, was developed to recognize different types of images. It showed good results on the MNIST and ORL databases. In this chapter,

Table 9.1 Results of system investigation

Threshold for specific point selection	Error number (four runs)	Average number of errors	% of correct recognition
20	3	3	85
	4		
	1		
	3		
40	2	2.25	88.75
	2		
	2		
	3		
60	3	1.5	92.5
	1		
	1		
	1		

the application of PCNC to micro workpiece shape recognition was investigated. The results show that this application of PCNC has a good perspective. Future work will involve online experiments using the neural recognition system on the base of micromachine tools and micromanipulators, which form the microfactory.

In 2006, we improved this result [14]. With a LIRA neural classifier, we obtained the recognition rate of 98.9%.

References

1. Kussul E., Rachkovskij D., Baidyk T. et al., Micromechanical Engineering: a Basis of the Low Cost Manufacturing of Mechanical Microdevices Using Microequipment. *Journal of Micromechanics and Microengineering*, 6, 1996, pp. 410–425.
2. Kussul E., Baidyk T., Ruiz-Huerta L., Caballero A., Velasco G., Kasatkina L. "Development of Micromachine Tool Prototypes for Microfactories", *Journal of Micromechanics and Microengineering*, 12, 2002, pp. 795–812.
3. Some Micro Machine Activities in Japan. Report ATIP96.021, 1996.
4. Okazaki Yuichi, Kitahara Tokio. Micro-Lathe Equipped with Closed-Loop Numerical Control. *Proceedings of the 2nd International Workshop on Microfactories*, Switzerland, Oct.9–10, 2000, pp. 87–90.
5. Naotake Ooyama, Shigeru Kokaji, Makoto Tanaka and others. Desktop Machining Microfactory. *Proceedings of the 2nd International Workshop on Microfactories*, Switzerland, Oct.9–10, 2000, pp. 14–17.
6. Bleuler H., Clavel R., Breguet J-M., Langen H., Pernette E. Issues in Precision Motion Control and Microhandling. *Proceedings of the 2000 IEEE International Conference On Robotics & Automation*, San Francisco, pp. 959–964.
7. Jonathan Wu Q.M., M.F. Ricky Lee, Clarence W. de Silva. Intelligent 3-D Sensing in Automated Manufacturing Processes. In: *Proceedings IEEE/ASME International Conference on Advanced Intelligent Mechatronics*, Italy, 2001, pp. 366–370.
8. Lee, S.J., K. Kim, D.-H. Kim, J.-O. Park, G.T. Park. Recognizing and Tracking of 3-D-Shaped Micro Parts Using Multiple Visions for Micromanipulation. In: *IEEE International Symposium on Micromechatronics and Human Science*, 2001, pp. 203–210.
9. Kim, J.Y., H.S. Cho. A Vision Based Error-Corrective Algorithm for Flexible Parts Assembly. In: *Proceedings of the IEEE International Symposium on Assembly and Task Planning*, Portugal, 1999, pp. 205–210.
10. Grigorescu C., Petkov N. Distance Sets for Shape Filters and Shape Recognition. IEEE Transactions on Image Processing, Vol. 12, No. 10, October 2003, pp. 1274–1286.
11. Kussul E.M., Baidyk T.N. Permutative Coding Technique for Handwritten Digit Recognition. *Proceedings of the 2003 IEEE International Joint Conference on Neural Networks (IJCNN'2003)*, Portland, Oregon, USA, July 20–24 2003, Vol. 3, pp. 2163–2168.
12. Kussul E., Baidyk T., Kussul M., 2004, Neural Network System for Face Recognition. 2004 IEEE International Symposium on Circuits and Systems, ISCAS 2004, May 23–26, Vancouver, Canada, Vol. V, pp. V-768–V-771.
13. Baidyk T., Kussul E., 2004, Neural Network Based Vision System for Micro Workpieces Manufacturing. *WSEAS Transactions on Systems*, Issue 2, Vol. 3, April 2004, pp.483–488.
14. Martín A., Baidyk T., Neural Classifier for Micro Screw Shape Recognition in Micromechanics, CLEI 2006, IFIP WCC AI2006, Santiago de Chile, Chile, August, 21–24 2006, p. 10.

Chapter 10
Texture Recognition in Micromechanics

The main approaches to microdevice production are microelectromechanical systems (MEMS) [1, 2] and microequipment technology (MET) [3–7]. To get the most out of these technologies, it is important to have advanced image recognition systems. In this chapter, we propose the Random Subspace Neural Classifier (RSC) for metal surface texture recognition.

10.1 Metal Surface Texture Recognition

Examples of metal surfaces are presented in Fig. 10.1. Due to changes in viewpoint and illumination, the visual appearance of different surfaces can vary greatly, making recognition very difficult [8]. Different lighting conditions and viewing angles greatly affect the grayscale properties of an image due to such effects as shading, shadowing, and local occlusions. The real surface images, which it is necessary to recognize in industrial environments, have all these problems and more, such as dust on the surface.

The task of metal surface texture recognition is important for automating the assembly processes in micromechanics [3]. To assemble any device, it is necessary to recognize the position and orientation of the workpieces to be assembled [4]. Identification of a workpiece surface is useful in recognizing its position and orientation. For example, let a shaft have two polished cylinder surfaces for bearings, one with a milled groove for a dowel joint, and the other turned by a lathe. It will be easier to obtain the orientation of the shaft if we can recognize both types of surface textures.

Our texture recognition system has the following structure (Fig. 10.2). The texture image serves as input data to the feature extractor. The extracted features are presented to the input of the encoder. The encoder produces an output binary vector of large dimension, which is presented to the input of the one-layer neural classifier. The output of the classifier gives the recognized class. Later, we will describe all these blocks in detail. To test our neural classifier RSC, we created our

E. Kussul et al., *Neural Networks and Micromechanics*,
DOI 10.1007/978-3-642-02535-8_10, © Springer-Verlag Berlin Heidelberg 2010

Fig. 10.1 Samples of metal surface after: a) sandpaper polishing, b) turning, c) milling, d) polishing with file

Feature Extractor → Encoder → Classifier

Fig. 10.2 Structure of RSC Neural Classifier

Fig. 10.3 Images of metal surface after milling

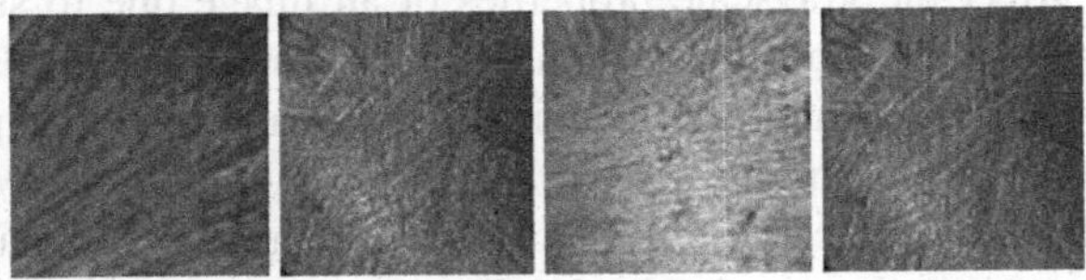

Fig. 10.4 Images of metal surface after polishing with sandpaper

own test set of metal surface images. We work with three texture classes, each containing 20 images. From these 20 images, we randomly select some for training the RSC, and the rest are used for testing. The number of images in the training set ranges from three to ten.

The first texture class corresponds to the metal surface after milling (Fig. 10.3), the second to the metal surface after polishing with sandpaper (Fig. 10.4), and the third to the metal surface after turning with a lathe (Fig. 10.5). You can see that different lighting conditions greatly affect the grayscale properties of an image. The texture may also be arbitrarily oriented, making the texture recognition task more complicated.

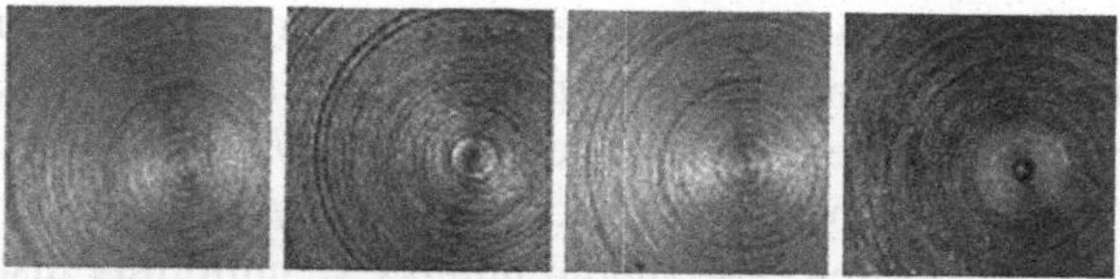

Fig. 10.5 Images of metal surface after turning with lathe

There is research on fast detection and classification of defects on treated metal surfaces using a back propagation neural network [9], but we do not know of any on texture recognition of metal surfaces after different mechanical treatments. Solving this problem can help us to recognize the positions and orientations of complex, mechanically processed workpieces.

The RSC neural classifier used in our system is based on the Random Threshold Classifier (RTC) developed earlier [10]. This section continues the series of works on automating micro assembly processes [3, 4]. The neural network classifier is developed to recognize textures of mechanically treated surfaces. This classifier can be used in recognition systems that have to recognize positions and orientations of complex workpieces in micromechanical device assembly tasks. To recognize the metal texture, we use the RSC (see Part 3).

10.2 Feature Extraction

Our image database for metal surface texture recognition consists of 60 grayscale images with a resolution of 220 x 220 pixels^2, 20 images for each of three classes. Image processing involves scanning across the initial image by moving a window of 40 x 40 pixels^2 with a step of 20 pixels. It is important to select the window size appropriately for all textures within the set to be classified because this gives us the opportunity to obtain local characteristics of the texture under recognition.

For every window, three histograms of brightness, contrast, and contour orientation were calculated. Every histogram contains 16 components, so we have the input feature vector containing 48 components. We present this input vector to our RSC classifier.

10.3 Encoder of Features

The encoder's task is to codify the feature vector $(X_1, \ldots, X_n)$ into binary form in order to present it to the input of the one-layer classifier. The feature encoder calculates the output values of the neurons $b_1, \ldots, b_s$.

To create the encoder structure, we have to select the subspaces for each neuron block. For example, if the subspace size is 3, in each neuron block j we will use only three input parameters whose numbers we select randomly from the range 1, . . ., n (where n is the dimension of the input space; in our case $n = 48$). After that, we calculate the thresholds for each pair of neurons l_i^j and h_i^j of three selected neurons a_i^j of block j. For this purpose, we select the point x_i^j randomly from the range [0,. . ., X_i]. After that, we select a random number y_i^j uniformly distributed in the range [0, . . . , GAP], where GAP is the parameter of the encoder structure. Then we calculate the thresholds of neurons l_i^j and h_i^j in accordance with formulas

$$\begin{aligned} &Trl_i^j = x_i^j - y_i^j; \\ &\text{if } (Trl_i^j < X_i min) \text{ then } Trl_i^j = X_i\, min; \end{aligned} \tag{10.1}$$

$$\begin{aligned} &Trh_i^j = x_i^j + y_i^j; \\ &\text{if}(Trh_i^j > X_i max) \text{ then } Trh_i^j = X_i max; \end{aligned} \tag{10.2}$$

where Trl_i^j and Trh_i^j are the thresholds of neurons l_i^j and h_i^jrespectively, and $X_i\, min$ and $X_i\, max$ are the minimum and maximum possible values for a component X_i of the input vector $(X_1, \ldots, X_n)$. Then the encoder forms a binary vector $(b_1, \ldots, b_s)$ corresponding to the feature vector. This vector is presented to the input of the one-layer classifier. The training rule of our one-layer classifier is the same as the training rule of the one-layer perceptron.

10.4 Results of Texture Recognition

We have conducted experiments with different numbers of images in training/test sets. Selecting for each texture class only three images for training and 17 images for recognition, we obtained a result of 80% correct recognition. The parameters of the RSC neural classifier were the following: number of b-neurons – 30,000; number of training cycles – 500.

There are some methods that work well when the features used for classifier training are obtained from a training sample that has the same orientation and position as the test sample, but as soon as the orientation and/or position of the test image changes with respect to the one in the training sample, the same methods will perform poorly. The usefulness of methods that are not robust to changes in orientation is very limited, which is why we developed our texture classification system that works well independently of the particular position and orientation of the texture. In this sense, the results obtained in experiments are sufficiently promising.

We train our RSC neural classifier with patterns in all the expected orientations and positions in such a way that the neural network becomes insensitive to them. The performance of the developed classifier was tested for recognition of three texture types obtained after milling, turning, and polishing of metal surfaces. The obtained recognition rate is 80%. In 2008, we improved this result with the LIRA neural classifier [11]. A recognition rate of 99.8% was obtained.

References

1. Trimmer W.S. (ed.), *Micromechanics and MEMS. Classical and seminal papers to 1990*, IEEE Press, New York, 1997.
2. Madni A.M., Wan L.A., Micro Electro Mechanical Systems (MEMS): An Overview of Current State-of-the Art. *Aerospace Conference*, Vol. 1, 1998, pp. 421–427.
3. Kussul E., Baidyk T., Ruiz-Huerta L., Caballero A., Velasco G., Kasatkina L., Development of Micromachine Tool Prototypes for Microfactories, *J. of Micromech. and Microeng.*, Vol. 12, 2002, pp. 795–813.
4. Baidyk T., Kussul E., Makeyev O., Caballero A., Ruiz L., Carrera G., Velasco G., Flat Image Recognition in the Process of Microdevice Assembly, *Pattern Recognition Letters*, Vol.25/1, 2004, pp. 107–118.
5. Frjedrich C.R., Vasile M.J., Development of the Micromilling Process for High-Aspect- Ratio Micro Structures, *J. Microelectromechanical Systems*, Vol.5, 1996, pp. 33–38.
6. Naotake Ooyama, Shigeru Kokaji, Makoto Tanaka, et al, Desktop Machining Microfactory, *Proceedings of the 2nd International Workshop on Microfactories*, Switzerland, 2000, pp. 14–17.
7. Baidyk T., Kussul E., Application of Neural Classifier for Flat Image Recognition in the Process of Microdevice Assembly, *IEEE Intern. Joint Conf. on Neural Networks*, Hawaii, USA, Vol. 1, 2002, pp. 160–164.
8. Matti Pietikäinen, Tomi Nurmela, Topi Mäenpää, Markus Turtinen, View-Based Recognition of Real-World Textures, *Pattern Recognition*, Vol.37, 2004, pp. 313–323.
9. Neubauer C., Fast Detection and Classification of Defects on Treated Metal Surfaces Using a Back Propagation Neural Network, *Proc. of the IEEE International Joint Conference on Neural Networks*, Vol. 2, 1991, pp. 1148–1153.
10. Kussul E., Baidyk T., Lukovitch V., Rachkovskij D., Adaptive High Performance Classifier Based on Random Threshold Neurons, *Proc. of Twelfth European Meeting on Cybernetics and Systems Research (EMCSR-94)*, Austria, Vienna, 1994, pp. 1687–1695.
11. Makeyev O., Sazonov E., Baidyk T., Martin A., Limited Receptive Area Neural Classifier for Texture Recognition of Mechanically Treated Metal Surfaces, *Neurocomputing*, Issue 7–9, Vol. 71, March 2008, pp. 1413–1421.

Chapter 11
Adaptive Algorithms Based on Technical Vision

We have developed technology for the automatic microassembly of microfilters – a description of the microfilters was presented in Chapter 8.

11.1 Microassembly Task

One problem with the microassembly process is that the workpiece sticks to the micromanipulator gripper, and it is difficult to release the gripper from the workpiece. To resolve this problem, we propose the following assembly process sequence [1] (Fig. 11.1). The gripper of the assembly device is the needle (1) in the tube (2). The microring is put on the needle and is introduced with the needle into the hole (Fig. 11.1a, b). After that, the needle is removed, and the microring is held in the hole with the tube (Fig. 11.1c). In the next step, the tube with the needle is moved aside, and the microring is held in the hole and cannot follow the tube (Fig. 11.1d). The tube then is moved up and liberates the end of the needle for the next operation (Fig. 11.1e).

Micromechanical device assembly requires the determination of the relative position of microdetails (Fig. 11.2). In the case of the pin-hole task, it is necessary to determine the displacements dX, dY, dZ of the pin tip relative to the hole by using the images obtained with a TV camera. It is possible to evaluate these displacements with a stereovision system, which resolves 3D problems but demands two TV cameras. To simplify the control system, we propose the transformation of 3D into 2D images, preserving all the information about the mutual location of the pin and the hole. This approach makes it possible to use only one TV camera.

Four light sources are used to obtain pin shadows (Fig. 11.3). The mutual location of these shadows and the hole contains all the information about displacements of the pin relative to the hole. The displacements in the horizontal plane (dX, dY) can be obtained directly by displacements of the shadows' points relative to the hole center. Vertical displacement of the pin may be obtained from the

DOI 10.1007/978-3-642-02535-8_11, © Springer-Verlag Berlin Heidelberg 2010

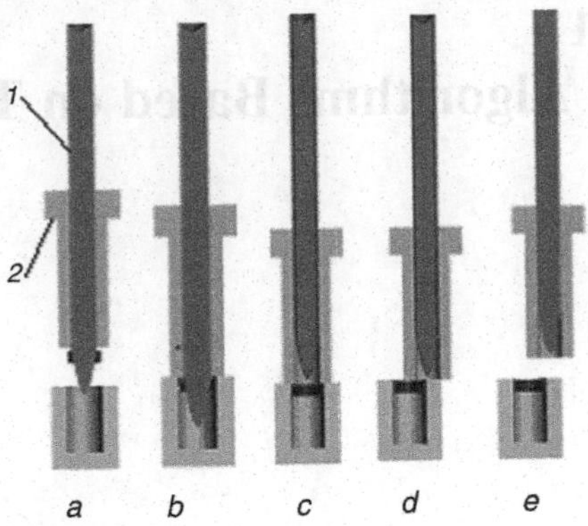

Fig. 11.1 The assembly process

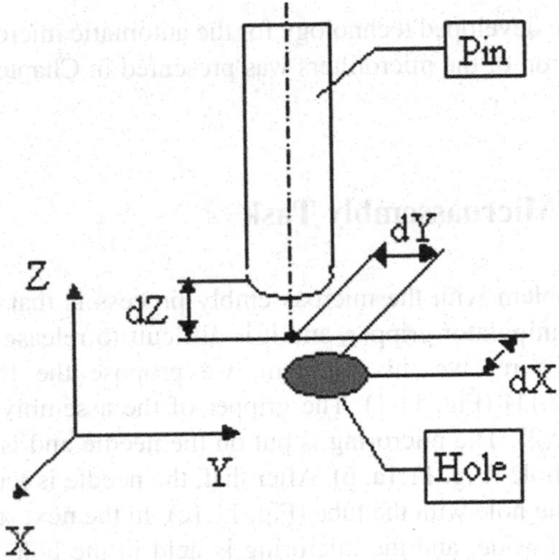

Fig. 11.2 Mutual location of the pin and hole

distance between the shadows. To calculate the displacements, it is necessary to have all the shadows in one image.

We capture four images corresponding to each light source, sequentially (Fig. 11.3), and then we extract contours and superpose contour images (Fig. 11.4). The image in Fig. 11.4 is similar to an optical character. It can be interpreted and treated as a symbol (resemblance to letters or numbers). For optical character recognition, we developed a Random Threshold Classifier and Random Subspace Classifier [2, 3] based on the modified Rosenblatt perceptron. We adapted the LIRA neural classifier to recognize the position of the pin [4 – 6], and we present its parameters below.

To examine the developed method of pin-hole position detection, we made two prototypes of the microassembly device (Fig. 11.5, Fig. 11.6). With these prototypes, we have measured only two coordinate displacements (*dX*, *dY*). We will do

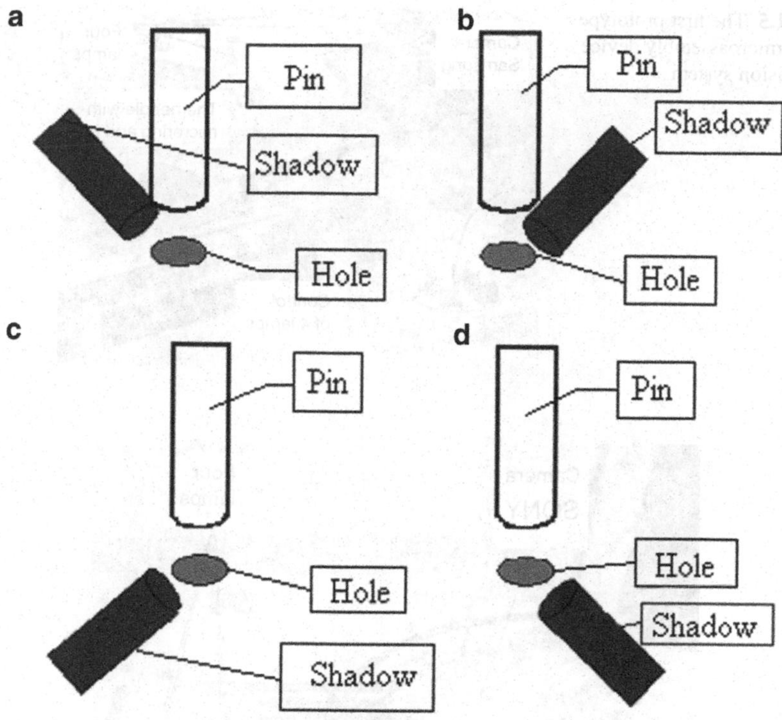

Fig. 11.3 Shadows from the pin that correspond to different light sources

Fig. 11.4 Contours of pin, hole, and shadows together

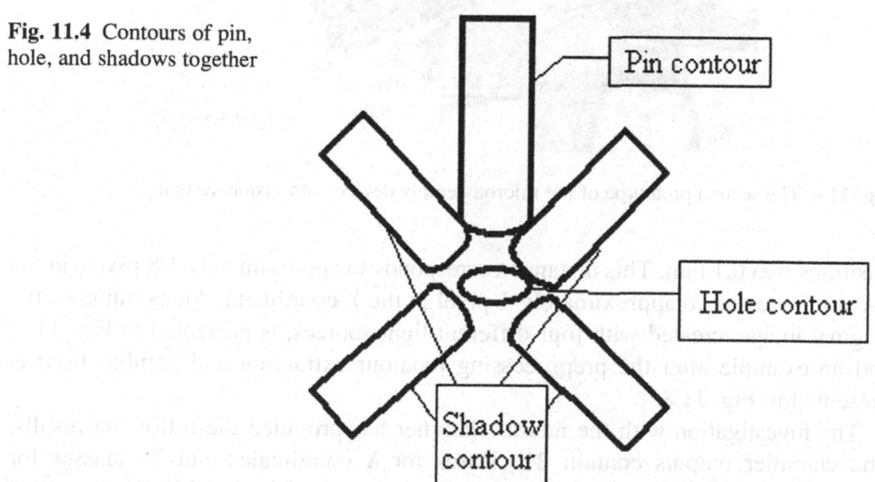

the experiments with the third coordinate *Z* in the future. With the second prototype, the images were taken from the 441 positions in the *X-Y* plane around the hole. The positions were located as a 21 x 21 matrix, and the distance between neighboring

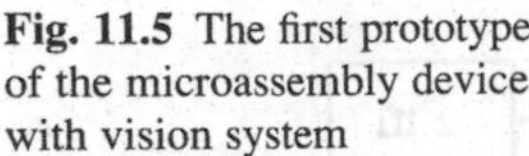

Fig. 11.5 The first prototype of the microassembly device with vision system

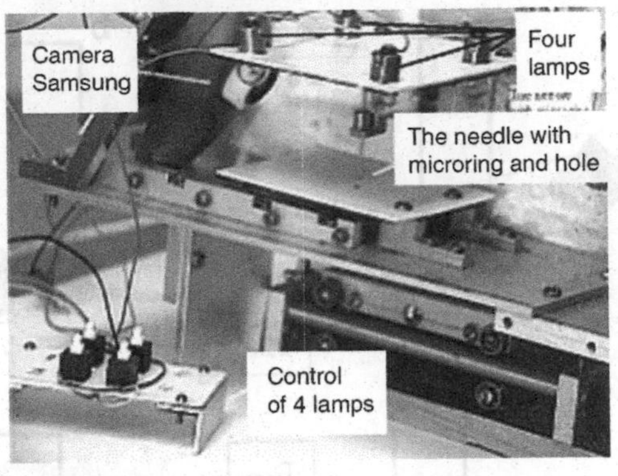

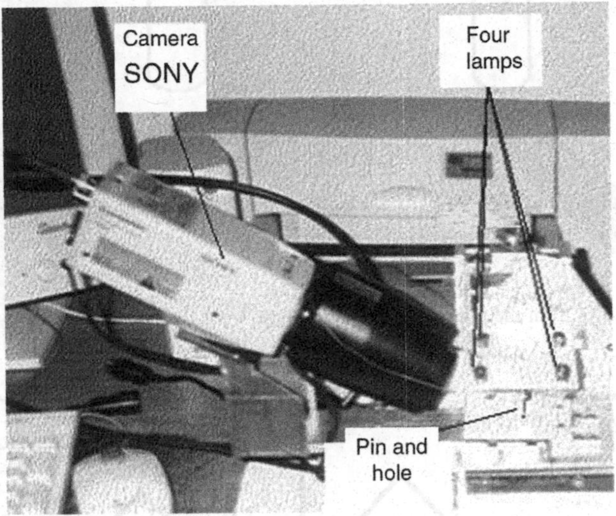

Fig. 11.6 The second prototype of the microassembly device with vision system

positions was 0.1 mm. This distance corresponds to approximately 1.8 pixels in the X-coordinate and to approximately 1 pixel in the Y-coordinate. An example of the original image, created with four different light sources, is presented in Fig. 11.7, and an example after the preprocessing (contour extraction and combination) is presented in Fig. 11.8.

The investigation with the neural classifier has provided the following results. The classifier outputs contain 21 classes for X coordinates and 21 classes for Y coordinates. The structure of the classifier is described in detail in [5, 6].

It is known [7] that the recognition rate may be increased if, during the training cycle, the images are represented not only in their initial state but also with shifted image positions (so-called distortions). We investigated a classifier with nine distortions. The training and the test sets were formed from 441 images. The

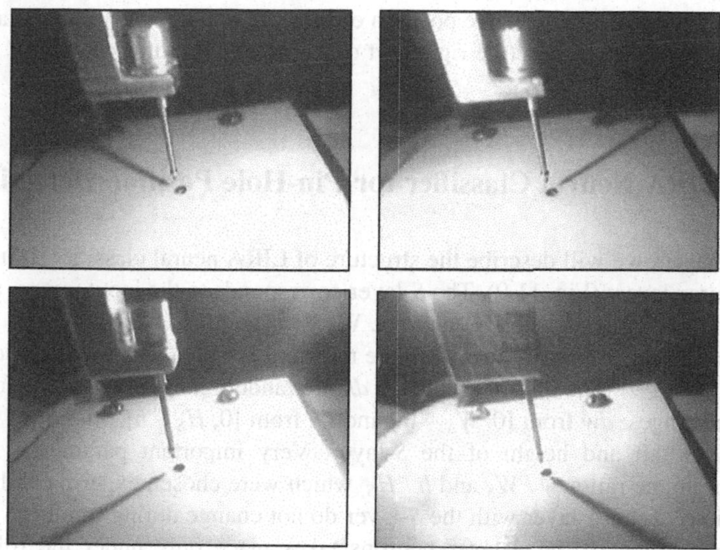

Fig. 11.7 Examples of the original images

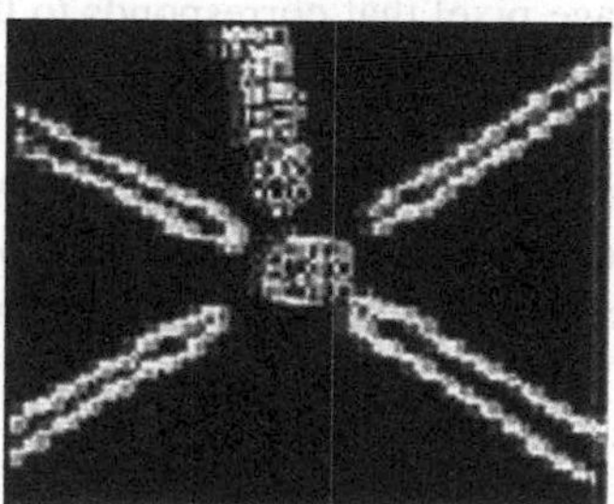

Fig. 11.8 Input image for neural network

Table 11.1 The results of the classifier investigation

X		Y		X±0.1mm		Y±0.1mm	
1	2 %	3	4 %	5	6 %	7	8 %
201.7	91.25	143.7	65.3	220	100	191.7	86.7

"chessboard" rule was used to divide the initial database into training and test datasets. The initial database was considered as a 21 x 21 chessboard. The "black" cells were considered as training samples and the "white" ones as test samples. In Table 11.1, the results of this investigation are presented [6]. The number of correct recognitions is presented in the first columns of coordinates *X* and *Y* (column numbers 1 and 3). Recognition rate (%) is in the second columns (column numbers

2 and 4). If we permit a pin-hole position error of up to 0.1 mm, we obtain a rather high quality of recognition (the right part of Table 11.1, columns 5 – 8).

11.2 LIRA Neural Classifier for Pin-Hole Position Detection

In this chapter, we will describe the structure of LIRA neural classifier. It includes three neuron layers (Fig. 11.9). The *S*-layer corresponds to the input image, and the *A*-layer is a layer of associative neurons. We connect each *A*-layer neuron with *S*-layer neurons randomly selected from the rectangle ($h * w$), which is located in the *S*-layer (Fig.11.9). The distances *dw* and *dh* are random numbers selected from the following ranges: *dw* from $[0, W_S - w]$ and *dh* from $[0, H_S - h]$, where W_S and H_S stand for width and height of the *S*-layer. Very important parameters of this classifier are the ratios w / W_S and h / H_S, which were chosen experimentally. The connections of the *A*-layer with the *S*-layer do not change during training.

The excitation of the *A*-layer neurons takes place only under the following condition. Every neuron of the *A*-layer has *m* positive connections and *l* negative connections with *S*-layer neurons. A positive connection is activated when the image pixel that corresponds to this neuron has a value of 1. A negative connection is activated when the image pixel that corresponds to this neuron has a value of 0. The excitation of the *A*-layer neuron takes place when all *m* positive connections and *l* negative connections are activated.

The *R*-layer presents the outputs of the classifier. All neurons of this layer are connected to all neurons of the *A*-layer with trainable connections. The excitation of *R*-layer neurons is defined in accordance with the formula:

$$E_j = \sum_{i=1}^{n} a_i w_{ij} \tag{11.1}$$

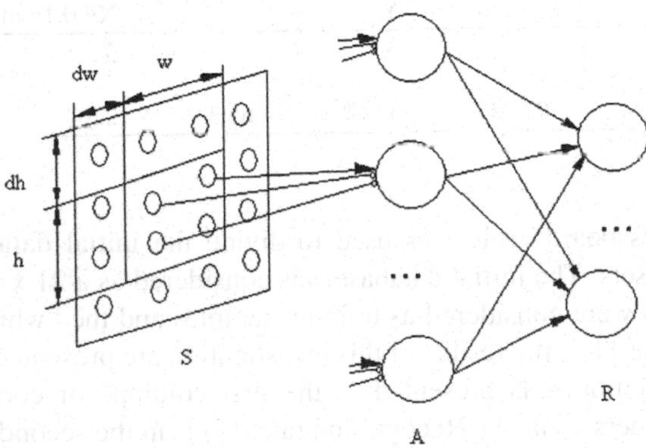

Fig. 11.9 LIRA Neural Classifier

where E_j is the excitation of j-neuron of the R-layer, a_i is the excitation of i-neuron of the A-layer, and w_{ij} is the connection weight between the A-layer neuron i and the R-layer neuron j.

In the neural classifier, the neuron from the R-layer having the highest excitation determines the class under recognition. This rule is used always in the process of recognition, but in training, it should be changed. Let the neuron-winner have excitation E_w and its nearest competitor have excitation E_c. If

$$(E_w - E_c)/E_w < T_E \tag{11.2}$$

the competitor is considered the winner. Here, T_E is the superfluous excitation of the neuron-winner.

Distinct from the Rosenblatt perceptron, LIRA neural classifier has only positive connections between the A-layer and the R-layer, so the training rule is as follows:

1. Let j correspond to the correct class under recognition. During recognition, we obtain excitations of R-layer neurons. The excitation of neuron R_j corresponding to the correct class is decreased by the factor (1 - T_E). After this, the neuron having maximum excitation R_k is selected as the winner.
2. If $j = k$, nothing must be done.
3. If $j \neq k$, $W_{ij}(t + 1) = W_{ij}(t) + a_i$, where $W_{ij}(t)$ is the weight of the connection between the i-neuron of the A-layer and the j-neuron of the R-layer before reinforcement, $W_{ij}(t + 1)$ is the weight after reinforcement, and a_i is the output signal (0 or 1) of the i-neuron of the A-layer.

$$W_{ik}(t+1) = W_{ik}(t) - a_i, \text{ if } (W_{ik}(t) > 0),$$
$$W_{ik}(t+1) = W_{ik}(t), \text{ if } (W_{ik}(t) = 0),$$

where $W_{ik}(t)$ is the weight of the connection between the i-neuron of the A-layer and the k-neuron of the R-layer before reinforcement, and $W_{ik}(t + 1)$ is the weight after reinforcement. According to this rule, connection weights have only positive values.

We adapt the classifier LIRA for pattern recognition in assembly of microdevices. The experiments were made with the pin, ring, and hole having the following parameters:

- diameter of the pin: 1 mm;
- outer diameter of the ring: 1.2 mm;
- inner diameter of the hole: 1.25 mm.

11.3 Neural Interpolator for Pin-Hole Position Detection

The neural interpolator [6, 8, 9] differs from the neural classifier because the excitations of the output neurons are considered as a set of values of continuous functions $f(dx)$ and $\varphi(dy)$. To determine the functions $f(dx)$ and $\varphi(dy)$, we use a parabolic regression equation based on five values selected from output neuron excitations according to the following rule. As a central point, we select the point with the maximal value of excitation E_{max} (Fig. 11.10a). In addition to this point, two points to the left and two points to the right of E_{max} are selected. If point E_{max} is located at the edge of the point sequence, additional points are obtained as a mirror reflection (Fig. 11.10b) of the points, which are situated on the other side of E_{max}. After determining the functions $f(dx)$ and $\varphi(dy)$, the parameters dx_0 and dy_0, under which the functions $f(dx)$ and $\varphi(dy)$ have maximal values, are determined. These parameters are recognized pin-hole displacements.

The interpolator training algorithm also differs from the classifier training algorithm, and the resulting training rule is as follows. Weight modification is carried out at every step according to the equation:

$$W_{ij}(t+1) = W_{ij}(t) + a_i * (\Delta w_j + \delta w_j), \tag{11.3}$$

where $W_{ij}(t)$ is the weight of the connection between the i-neuron of the A-layer and the j-neuron of the R-layer before reinforcement, $W_{ij}(t + 1)$ is the weight after reinforcement, and a_i is the output signal (0 or 1) of the i-neuron of the A-layer.

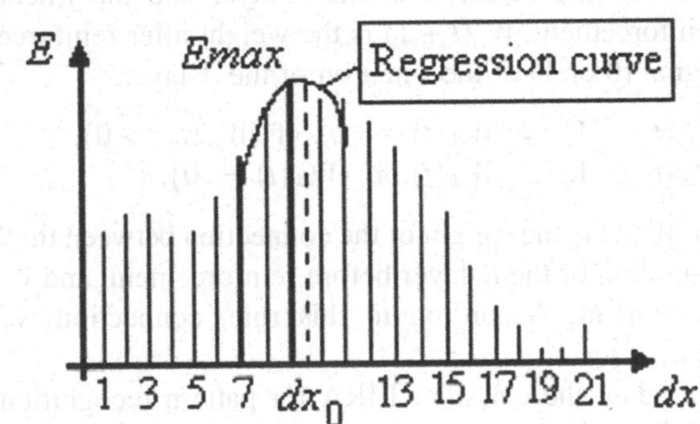

Fig. 11.10a Parabolic approximation

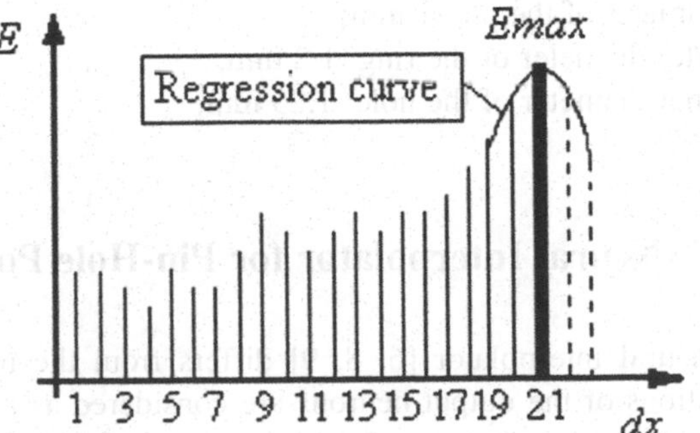

Fig. 11.10b Parabolic approximation

$$\Delta w_j = \frac{1}{(dx_j - dx_c)^2 + \varepsilon},$$
$$\delta w_j = -\frac{1}{(dx_j - dx_O)^2 + \varepsilon}, \tag{11.4}$$

where dx_c is the correct pin-hole displacement, dx_O is the recognized pin-hole displacement, Δw_j and δw_j are the positive and the negative training factors for the *j*th neuron, and dx_j is the pin-hole displacement corresponding to the *j*th neuron (Fig. 11.11). For coordinate *Y*, similar formulas are used.

The results of the investigation of the neural interpolator are shown in Table 11.2 for coordinate *X* and Table 11.3 for coordinate *Y*. In Tables 11.1, 11.2, and 11.3, the average results for six independent experiments are presented. The experiments show that the computer vision system can recognize a relative pin-hole position with a 0.1 mm tolerance. The neural classifier for this tolerance gives the correct recognition in 100% of cases for *X* and 86.7% of cases for *Y* (Table 11.1). The neural interpolator gives 100% for axis *X* (Table 11.2) and 99.86% for axis *Y* (Table 11.3). The neural interpolator also permits us to obtain data for smaller tolerances. For example, for axis *X* with 0.05 mm tolerance, it gives an 88.6% recognition rate, and for axis *Y* with 0.05 mm tolerance, it gives a 79.1% recognition rate. The experiments show that the neural interpolator gives better results in estimating pin-hole relative positions.

It is interesting to observe that the 0.05 mm tolerance for the *Y* axis is less than one pixel in the image. In this case, the recognition rate of 79.1% shows that recognition possibilities are not limited by the resolution of one pixel. This result could be explained by the fact that each object in the image contains many pixels, which give much more detailed information than one pixel.

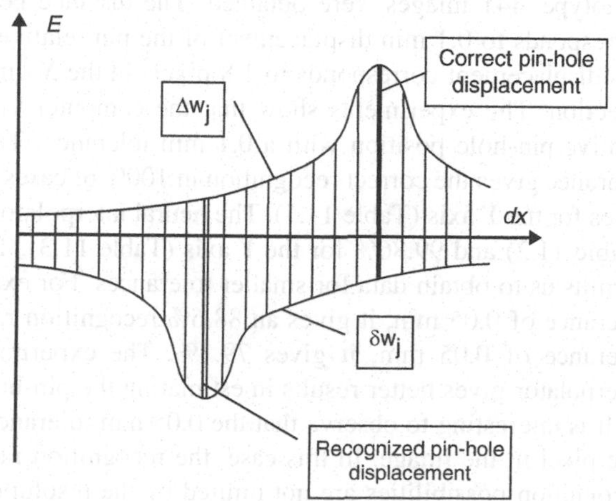

Fig. 11.11 Correct and recognized displacements

Table 11.2. The results for *X* coordinate of interpolator investigation

ε for recognition	X±0.025 mm		X±0.05 mm		X±0.075 mm		X±0.1mm		X±0.15mm	
	X	X%	X	X%	X	X%	X	X%	X	X%
Error	74	33.6	25	11.4	4	1.8	0	0	0	0
Correct recognition	146	66.4	195	88.6	216	98.2	220	100	220	100

Table 11.3. The results for *Y* coordinate of interpolator investigation

ε for recognition	Y± 0.025 mm		Y± 0.05 mm		Y± 0.075 mm		Y± 0.1 mm		Y± 0.15 mm	
	Y	Y%	Y	Y%	Y	Y%	Y	Y%	Y	Y%
Error	101	45.9	48	20.9	11	5	0.3	0.14	0	0
Correct recognition	119	54.1	172	79.1	209	95	219.7	99.86	220	100

11.4 Discussion

The equipment for automatic assembly of microdevices must have high precision because the tolerances of microdevice components are very small. To increase precision, we use feedback based on computer vision principles. We propose an image recognition method based on neural networks and have developed two types of neural networks to solve this problem [4, 6, 10, 11]. We call the first type a neural classifier and the second a neural interpolator. They are used to obtain relative positions of the micropin and microhole. The neural classifier gives a finite number of relative positions, while the neural interpolator serves as an interpolation system and gives an infinite number. An image database of relative positions of the microhole and micropin of diameter 1.2 mm was used to test the neural classifier and the neural interpolator. In this work, the recognition results of the relative positions are presented for the neural classifier and the neural interpolator.

A special prototype was made to examine the proposed method. With this prototype, 441 images were obtained. The distance between neighboring images corresponds to 0.1 mm displacement of the pin relative to the hole. At the image, this displacement corresponds to 1.8 pixels in the *X* direction and 1 pixel in the *Y* direction. The experiments show that the computer vision system can recognize relative pin-hole position with a 0.1 mm tolerance. The neural classifier for this tolerance gives the correct recognition in 100% of cases for the *X* axis and 86.7% of cases for the *Y* axis (Table 11.1). The neural interpolator gives 100% for the *X* axis (Table 11.2) and 99.86% for the *Y* axis (Table 11.3). The neural interpolator also permits us to obtain data for smaller tolerances. For example, for an *X* axis with a tolerance of 0.05 mm, it gives an 88.6% recognition rate, and for a *Y* axis with a tolerance of 0.05 mm, it gives 79.1%. The experiments show that the neural interpolator gives better results in estimating the pin-hole relative positions.

It is interesting to observe that the 0.05 mm tolerance for the *Y* axis is less than one pixel in the image. In this case, the recognition rate of 79.1% shows that the recognition possibilities are not limited by the resolution of one pixel. This result could be explained by the fact that each object in the image contains many pixels and these pixels give much more detailed information than one pixel.

A computer vision system for improving the microassembly device's precision is proposed. Two algorithms of image recognition (neural classifier and neural interpolator) were tested in the task of pin-hole relative position detection. The neural classifier permits us to recognize the displacement of the pin relative to the hole with 1 pixel tolerance, while the neural interpolator permits recognition with 0.5 pixel tolerance. The absolute values of detectable displacements depend on the optical channel resolution. In our case, one pixel corresponds to 0.05 mm X-axis displacements and 0.1 mm Y-axis displacements. This precision is sufficient for many different assembly processes.

References

1. Kussul E., Baidyk T., Ruiz-Huerta L., Caballero A., Velasco G., Kasatkina L., 2002, Development of Micromachine Tool Prototypes for Microfactories, *Journal of Micromechanics and Microengineering*, 12, pp. 795–813.
2. Kussul E., Baidyk T. Neural Random Threshold Classifier in OCR Application. In: Proceedings of the Second All-Ukrainian International Conference, Kiev, Ukraine, 1994, pp. 154–157.
3. Kussul E., Baidyk T., Lukovitch V., Rachkovskij D. Adaptive High Performance Classifier Based on Random Threshold Neurons. In R. Trappl (ed.) Cybernetics and Systems'94, Singapore: World Scientific Publishing Co. Pte. Ltd., 1994, pp. 1687–1695.
4. Baidyk T. Application of Flat Image Recognition Technique for Automation of Micro Device Production. In: Proceedings of 2001 IEEE/ASME International Conference on Advanced Intelligent Mechatronics, Teatro Sociale Como, Italy, 2001, pp. 488–494.
5. Baidyk T., Kussul E. Application of Neural Classifier for Flat Image Recognition in the Process of Microdevice Assembly. In: Proceedings of IEEE International Joint Conference on Neural Networks, Hawaii, USA, Vol. 1, 2002, pp. 160–164.
6. Baidyk T., Kussul E., Makeyev O., Caballero A., Ruiz L., Carrera G., Velasco G. Flat Image Recognition in the Process of Microdevice Assembly. *Pattern Recognition Letters*, Vol. 25/1, 2004, pp. 107–118.
7. Bottou L., Cortes C., Denker J., Drucker H., Guyon L., Jackel L., LeCun Y., Muller U., Sackinger E., Simard P., Vapnik V., Comparison of Classifier Methods: A Case Study in Handwritten Digit Recognition, Proc. of 12th IAPR Intern. Conf. on Pattern Recognition, 1994, Vol. 2, pp. 77–82.
8. Makeyev O., Neural Interpolator for Image Recognition in the Process of Microdevice Assembly. In: IEEE IJCNN'2003, Portland, Oregon, USA, July 20–24, Vol. 3, 2003, pp. 2222–2226.
9. Baidyk T., Kussul E., Makeyev O. Image Recognition System for Microdevice Assembly, Twenty-First IASTED International Multi-Conference on Applied Informatics, AI2003, Innsbruck, Austria, 2003, pp. 243–248.
10. Kussul E., Baidyk T., 2004, Improved Method of Handwritten Digit Recognition Tested on MNIST Database. *Image and Vision Computing*, Vol. 22/12, pp. 971–981.
11. Kussul E., Baidyk T., 2006, LIRA Neural Classifier for Handwritten Digit Recognition and Visual Controlled Microassembly. *Neurocomputing*, Vol. 69, Issue 16–18, 2006, pp. 2227–2235.